Basic Numerical Methods in Meteorology and Oceanography

Kristofer Döös, Peter Lundberg & Aitor Aldama Campino

Published by
Stockholm University Press
Stockholm University
SE-106 91 Stockholm
Sweden
https://www.stockholmuniversitypress.se/

ORCiD:
Kristofer Döös: 0000-0002-1309-5921
Peter Lundberg: 0000-0002-6832-9836
Aitor Aldama Campino: 0000-0001-8453-4322

First published 2022
Cover designed by Sofie Wennström, Stockholm University Press

ISBN (Paperback): 978-91-7635-175-8
ISBN (PDF): 978-91-7635-172-7
ISBN (EPUB): 978-91-7635-173-4
ISBN (Mobi): 978-91-7635-174-1

DOI: https://doi.org/10.16993/bbs

Suggested citation:
Döös, K., Lundberg, P., and Campino, A. A. 2022. *Basic Numerical Methods
in Meteorology and Oceanography*. Stockholm: Stockholm University Press. DOI:
https://doi.org/10.16993/bbs. License: CC BY 4.0.

To read the free, open access version of this book online,
visit https://doi.org/10.16993/bbs or scan this QR code
with your mobile device.

Editorial Board for the book Basic Numerical Methods in Meteorology and Oceanography

The role of an Editorial Board working with book projects at Stockholm University Press is to be directly involved with the editorial quality assurance process, meaning that they read and comment on the book in all stages as well as suggest reviewers to invite. Their role is to ensure that the peer-review process is to be trusted and that the author receives constructive feedback in order to improve the academic quality of the book.

The book *Basic Numerical Methods in Meteorology and Oceanography* is not included in any of the book series produced by Stockholm University Press, and the Editorial Board is, therefore, specially invited to contribute with their expertise to this particular project.

The responsibilities of an Editorial Board at Stockholm University Press include:

- Making academic assessments of incoming book proposals and engaging with the peer-review process for book proposals and book manuscripts.
- Aiming to improve the academic quality of the material sent for evaluation through a peer-review process and adhering to good editorial practices according standards defined by the Committee of Publication Ethics (COPE, https://publicationethics.org).
- Championing the freedom of expression.
- Maintaining the integrity of the academic record.
- Ensuring that the intellectual and ethical standards are precluding any business or institutional needs.
- Ensuring that the published books maintain high standards for the refereeing and acceptance of the final version.

The Editorial Board consisted of the following researchers:

Peer Review Policy

Stockholm University Press ensures that all book publications are peer-reviewed. Each proposal submitted to the Press will be sent to a dedicated Editorial Board of experts in the subject area for evaluation. The full manuscript will be reviewed by chapter or as a whole by at least two external and independent experts.

A complete description of Stockholm University Press' peer-review policies can be found on the website: https://www.stockholmuniversitypress.se/site/peer-review-policies/

The Editorial Board of *Basic Numerical Methods in Meteorology and Oceanography* used an external anonymized peer-review procedure while evaluating the book proposal as well as the final book manuscript was. The Editorial Board and Stockholm University Press expresses its sincere gratitude towards all researchers involved in this project.

Recognition for reviewers

The Publisher and the Editorial Board would like to extend a special thanks to the reviewers who contributed to the process of editing this book.

Both the book proposal and the final manuscript was reviewed by the following external experts:

Heikki Järvinen, Professor, Institute for Atmospheric and Earth System Research, Helsinki University, Finland
ORCiD: https://orcid.org/0000 0003-1879-6804

Eigil Kaas, Professor, Physics of Ice, Climate and Earth, University of Copenhagen, Denmark
ORCiD: https://orcid.org/0000-0001-6970-2404

Contents

Preface

The purpose of this book is to provide an introduction to and an overview of numerical modelling of the ocean and the atmosphere. It has evolved from a course given at Stockholm University since 1997 and is intended to serve as a textbook for students in meteorology and oceanography at the master level. A prerequisite is a background in mathematics, physics, some geophysical fluid dynamics and programming. Focus will be on numerical schemes for the most common equations in oceanography and meteorology as well as on the stability, precision and other basic numerical properties of these schemes. We will use as simple equations as possible that still capture the properties of the primitive equations used in the general circulation models. For simplicity, the equations will often be referred to as the *hydrodynamic* equations since the numerical methods to be described here are valid for modelling both the ocean and the atmosphere. Due to the non-linearity of these equations, it is not possible to find analytical solutions. The equations are therefore instead solved numerically on a grid by discretisation, and the derivatives of the differential equations are replaced by finite-difference approximations. This is what constitutes a numerical model, which often is referred to as a general circulation model when it represents the three-dimensional (3D) global circulation of the atmosphere or the ocean. These models are based on the Navier-Stokes equations (including the Coriolis effect) and a tracer equation for heat in both the atmosphere and ocean and tracer equations for humidity and salt in the atmosphere and ocean, respectively. A coupled atmospheric and oceanic general circulation model represents the core part of an Earth System climate model.

The focus here will be on the basic numerical methods used for oceanographic and atmospheric modelling. For more detailed and comprehensive books see *e.g.* Durran (2010) for the numerical methods and *e.g.* Kalnay (2003) for data assimilation and numerical weather prediction.

The book starts with a short summary of the historical background of the numerical modelling of the ocean and atmosphere. In Chapter 2, we present the most common types of Partial Differential Equations (PDEs) that occur in meteorology and oceanography and how they can be classified. Chapter 3 introduces some of the most commonly used finite differences in both time and space and how accurate they are compared to the derivative. In Chapter 4, we show how to undertake a stability analysis using, for simplicity, the advection equation. The stability is shown to depend both on which finite-difference schemes that are used and on the advection speed, time step and grid size. Some schemes turn out to be unconditionally unstable, but can still be used as a first time step. In Chapter 5 we show how, in addition to the physical mode, a numerical mode arises and how this mode can be damped or suppressed. The accuracy of the numerical phase speed is examined in Chapter 6 and how this depends on the wave number and grid resolution. The shallow-water equations are discretised in Chapter 7 using three different spatial grids. Many limitations, which previously have been studied for the advection equation, are here examined again, but for the shallow-water equations. In Chapter 8 we investigate the discretisation of the friction and the diffusion terms and how this affects the stability of the solution. Iterative methods for the Poisson and Laplace equations are demonstrated in Chapter 9. Implicit and semi-implicit schemes are shown in Chapter 10. Chapter 11 deals with the semi-Lagrangian technique for the advection equation. Different model coordinates for atmospheric as well as oceanic models are presented in Chapter 12. Using a highly simplified approach, 3D-modelling is introduced in Chapter 13. Chapter 14 gives a brief description of how some atmospheric general circulation models use spectral methods as "horizontal coordinates". Some "pen-and-paper" theoretical exercises are given in Chapter 15 and Chapter 16 is devoted to a number of GFD computer exercises.

We wish to thank Laurent Brodeau, Bror Jönsson, Joakim Kjellsson, Gurvan Madec as well as our students for valuable input during all these years.

Kristofer Döös, professor of climate modelling,
Peter Lundberg, professor emeritus of physical oceanography
and Dr. Aitor Aldama Campino, all at the Department of
Meteorology, Stockholm University

1. Introduction

1.1 What is a numerical model of the circulation of the atmosphere or the ocean?

A numerical model of the circulation of the atmosphere and/or the ocean is basically constituted by

- A grid covering the spatial domain under consideration.
- Discretised equations describing the conservation of momentum, mass, energy and salt/moisture.
- Open and/or solid boundaries.
- A specified initial state of the system.

The schematic cartoon shown in Figure 1.1 illustrates a system of this type pertaining to a coupled ocean-atmosphere model with corresponding forcing and the exchange between the two components of the model.

1.2 Brief historical background

A short summary is here given of the historical background to the numerical modelling of the ocean and atmosphere.

Bjerknes (1904) was the first to discuss the possibility of predicting and modelling the circulation of the atmosphere. For this purpose he proposed seven equations with seven unknown variables:

- Three equations of conservation of momentum for the three velocity components based on Newton's second law.
- The continuity equation, *i.e.* the conservation of mass.

How to cite this book chapter:
Döös, K., Lundberg, P., and Campino, A. A. 2022. *Basic Numerical Methods in Meteorology and Oceanography*, pp. 1–5. Stockholm: Stockholm University Press.
DOI: https://doi.org/10.16993/bbs.a. License: CC BY 4.0

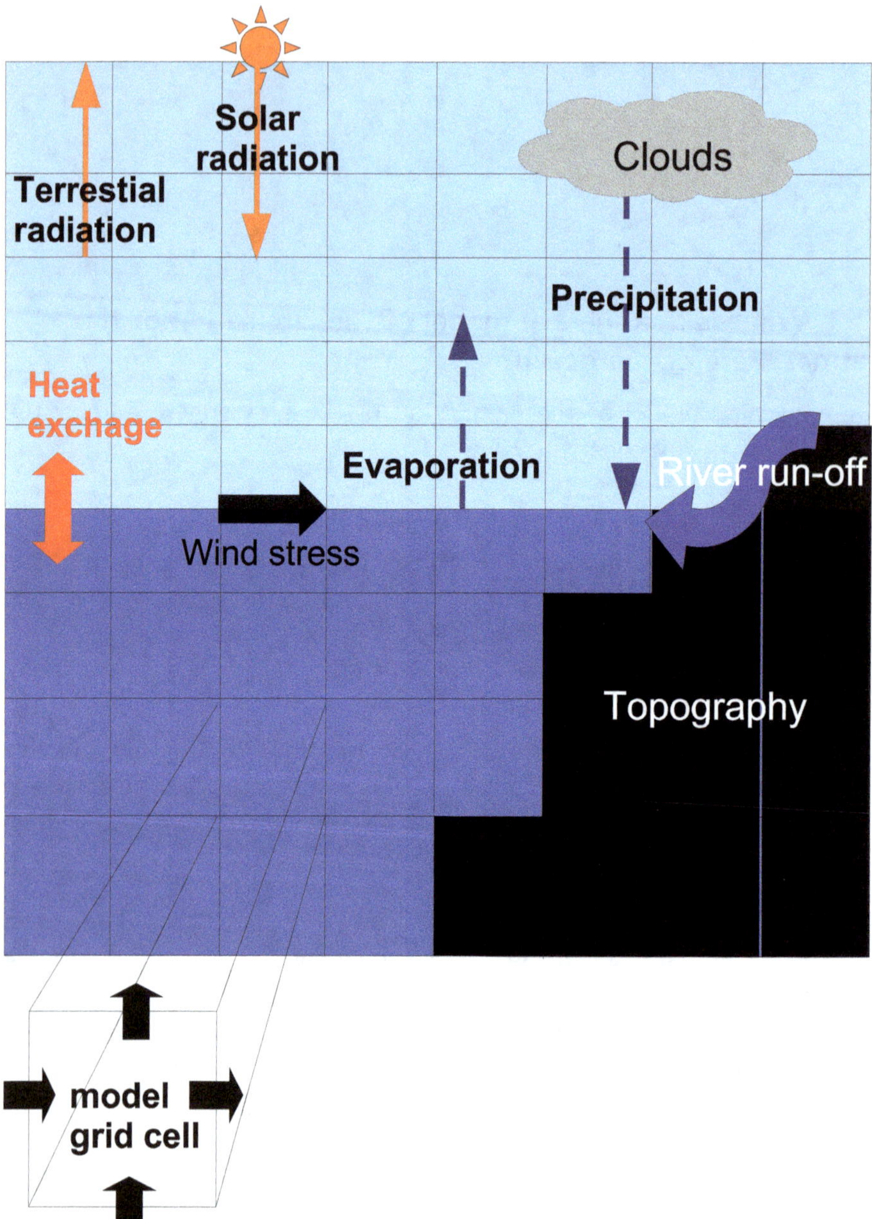

Figure 1.1. Schematic illustration of a numerical ocean-atmosphere circulation model

- The equation of state for ideal gases.
- The equation of conservation of energy based on the first law of thermodynamics.
- A conservation equation for water mass in the atmosphere.

These equations have become known as the "primitive equations" since they do not deal with a filtered quasi-geostrophic equation but go back to Bjerknes' original formulation of the problem. These equations also require boundary conditions at the top and bottom of the atmosphere as well as the initial state of the atmosphere. The problem is that although we have an equal number of equations and variables, as well as appropriate boundary and initial conditions, these equations cannot be solved analytically.

The first attempt to solve the equations numerically was made by Lewis Fry Richardson in his spare time when serving as an ambulance driver during the First World War. Richardson (1922) applied finite-differencing techniques to the equations suggested by Bjerknes.

A decade earlier Richardson (1911) had been the first to propose the use of finite differences when approximating partial differential equations, insights he put to use when attempting numerical weather prediction. His first numerical weather prediction (NWP) failed, however, due to two reasons. First, no computers were available and Richardson estimated that it would require a staff of 64,000 persons to compute a 24-hour forecast in 24 hours. The second reason, which was revealed by Lynch (2006), was due to a failure to apply smoothing techniques to the data, which led to his calculations yielding erroneous results.

These numerical instabilities were independently investigated by Courant, Friedrichs and Lewy (Courant et al., 1928), pure mathematicians who examined techniques of solving general partial differential equations using finite differencing. They found certain conditions regarding the choice of time step and grid size that must be respected for the numerical solution to be stable. These results were later further developed by Charney, Fjørtoft and von Neumann (Charney et al., 1950), the trio that in the late 1940s made the first successful numerical weather forecast, based on integration of the conservation equation for the absolute vorticity.

This single-equation approach of following the motion of an air column instead of using the seven-equation set proposed by Bjerknes (1904) was suggested by Carl-Gustaf Rossby, who was also the driving force behind the first real-time numerical weather prediction made in two runs 23-24 March 1954 in Stockholm by Harold Bedient and Bo Döös (Döös and Eaton, 1957; Wiin-Nielsen, 1991; Persson, 2005a).

The rationale underlying the need of numerical weather prediction is immediately obvious; less so when ocean forecasting is concerned, where two severe inundations of the northwestern European coast served as the driving agent. In the Netherlands, severe flooding in 1916 with around 300 fatalities led the government to entrust the physicist and Nobel Laureate Hendrik Lorentz with the task of developing predictive techniques. This ultimately led to Lorentz formulating a method for storm-surge forecasting in the 1920s. An even worse instance of North-Sea coastal flooding took place in 1953, with around 2000 deaths, which led to the Hamburg oceanographer Hansen (1956) developing a numerical storm-surge model based on finite differencing of the shallow-water equations including the non-linear advection terms. Hansen focused on the fast barotropic gravity waves instead of filtering them out as was done in numerical weather-prediction models. Subsequently another German oceanographer (Fischer, 1959) constructed a numerical shallow-water model for the North Sea with the finite-difference schemes described in detail.

The return to the full primitive equations for NWP models, which Richardson (1922) had used in the very beginning, was inevitable since the quasi-geostrophic equations, although very useful for understanding the large-scale extratropical dynamics of the atmosphere, were not accurate enough to allow continued progress in NWP. A complete Atmospheric General Circulation Model (AGCM) based on the primitive equations was developed by Smagorinsky (1963) at the Geophysical Fluid Dynamics Laboratory (GFDL) in Princeton. Here also a large part of the subsequent model development took place, including the first Ocean General Circulation Model (OGCM), developed only a few years later by Bryan and Cox (1967). Syukuro Manabe and Kirk Bryan combined their models to yield the first coupled Atmosphere-Ocean General Circulation Model (AOGCM), cf. Manabe and Bryan (1969), which they used for climate studies.

Ocean and atmosphere GCMs have since then increased in number all over the world and have improved in many aspects; higher resolution using more powerful computers, better parameterisations, more observations to feed the models with. The basic numerics have also improved, but the fundamental properties of the finite differences and the numerical methods remain. This is why focus in this book will be on the limitations of the numerical schemes that to a large extent were discovered during the 20th century.

For a more comprehensive view of the historical background, there are number of studies focusing on the evolution of NWP, *e.g.* Persson (2005a,b,c) and Wiin-Nielsen (1991).

2. Partial Differential Equations

Before going into the details of numerical procedures, we first classify the most usual types of Partial Differential Equations (PDEs) that occur in meteorology and oceanography. Partial differential equations are of vast importance in applied mathematics and engineering since so many real physical processess can be modelled by them. Second-order linear PDEs of the type needed for modelling the ocean and atmosphere circulation can be classified into three categories: elliptic, parabolic and hyperbolic. These partial differential equations in two dimensions are general linear equations of second order:

$$a\frac{\partial^2 u}{\partial x^2} + b\frac{\partial^2 u}{\partial x \partial y} + c\frac{\partial^2 u}{\partial y^2} + d\frac{\partial u}{\partial x} + e\frac{\partial u}{\partial y} + fu + g = 0. \qquad (2.1)$$

This equation has two independent variables (x and y) and one dependent (u) as well as two unspecified functions (f and g). We will show here that by transforming x, y and u into new variables, it is possible to apply Equation 2.1 to any second-order linear PDE and to classify it.

As the reader may note, this expression bears somewhat of a resemblance to the equation for a conic section:

$$ax^2 + bxy + cy^2 + dx + ey + f = 0, \qquad (2.2)$$

where a, b, c, d, e and f are constants. As illustrated by Figure 2.1, this algebraic equation represents an ellipse, parabola or a hyperbola depending on whether $b^2 - 4ac$ is negative, equal to zero or positive, respectively.

This indicates that one analogously can classify the PDE according to Table 2.1.

How to cite this book chapter:
Döös, K., Lundberg, P., and Campino, A. A. 2022. *Basic Numerical Methods in Meteorology and Oceanography*, pp. 7–12. Stockholm: Stockholm University Press. DOI: https://doi.org/10.16993/bbs.b. License: CC BY 4.0

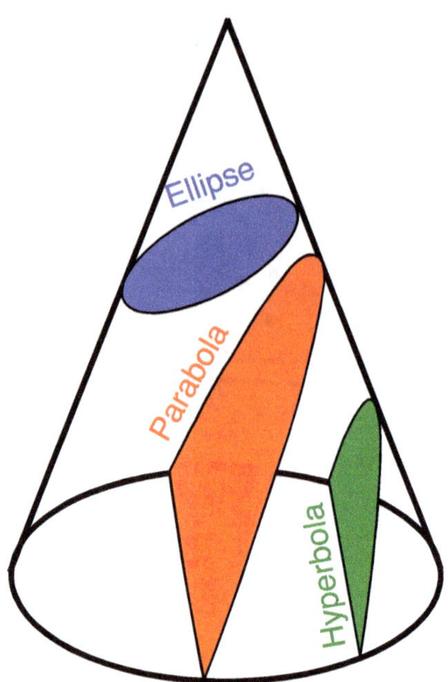

Figure 2.1. A diagram showing conic sections, *viz.* an ellipse, a parabola and a hyperbola. Note that these geometrical representations of Equation (2.2) are not solutions to the general PDE given by Equation (2.1).

Table 2.1. Classification of the three PDE types and their different kinds of boundary conditions. A Dirichlet boundary condition specifies the function on the boundary of the domain. A Neumann boundary condition specifies the value of the normal derivative of the function on the boundary of the domain. A Cauchy boundary condition specifies the values of the function and its normal derivative on the boundary of the domain.

PDE	$b^2 - 4ac$	Boundary & initial conditions	Examples
Elliptic	$b^2 - 4ac < 0$	Dirichlet/Neumann/Cauchy	Laplace Eq.
Parabolic	$b^2 - 4ac = 0$	One initial + two boundary conditions	Diffusion Eq.
Hyperbolic	$b^2 - 4ac > 0$	Two initial + two boundary conditions	Wave Eq.

Each type of system is associated with significantly different characteristic behaviour, and the solution scheme for each type of equation can also differ. All three classes of PDEs are represented among the most common equations in hydrodynamics and require the specification of different kinds of boundary conditions. We will now provide some examples of these equations.

2.1 Elliptic equations

The two-dimensional (2D) versions of the Laplace and Poisson equations are classical examples of elliptic equations, representing *e.g.* the relationship between the stream function and vorticity or the steady-state temperature in a plate:

$$\frac{\partial^2 u}{\partial x^2} + \frac{\partial^2 u}{\partial y^2} = 0 \ (or \ g\,(u, x, y)). \tag{2.3}$$

In conformity with Equation (2.1) it is recognised that $a = 1$, $b = 0$, $c = 1$ and $d = e = f = 0$, which leads to $b^2 - 4ac = -4 < 0$, and hence the PDE must be elliptic.

Another example of an elliptic equation in hydrodynamics can be found from the simplest equations of motion in the atmosphere and the ocean, *viz.* the linearised shallow-water equations without friction. These equations can be both hyperbolic (as shown in Section 2.3) and elliptic when in steady state as will be shown here.

$$\frac{\partial u}{\partial t} - fv = -g\frac{\partial \eta}{\partial x}, \tag{2.4a}$$

$$\frac{\partial v}{\partial t} + fu = -g\frac{\partial \eta}{\partial y}, \tag{2.4b}$$

$$\frac{\partial h}{\partial t} + H\left(\frac{\partial u}{\partial x} + \frac{\partial v}{\partial y}\right) = 0. \tag{2.4c}$$

Here u and v are the velocity components in the x and y direction, respectively, η is the surface-height perturbation from the undisturbed depth H of the fluid, g is the acceleration of gravity and t represents the time. Above $f \equiv 2\Omega \sin\varphi$ is the Coriolis acceleration, where Ω is the angular frequency of the Earth's rotation and φ the latitude. In what follows f is set to be a constant.

Note that the variables x and y in Equation (2.1) have been designated in an arbitrary fashion and do not have anything to do with the independent variables x and y in Equations (2.4). From these equations it can be deduced that

$$\frac{\partial}{\partial t}\left[\frac{\partial^2}{\partial t^2} + f^2 - gH\left(\frac{\partial^2}{\partial x^2} + \frac{\partial^2}{\partial y^2}\right)\right]\eta = 0. \tag{2.5}$$

If we integrate this equation in time we obtain

$$\frac{\partial^2 \eta}{\partial t^2} + f^2 \eta - gH \left(\frac{\partial^2}{\partial x^2} + \frac{\partial^2}{\partial y^2} \right) \eta = \xi\,(x, y), \qquad (2.6)$$

where $\xi(x, y)$ is an integration constant independent of t. When the problem is stationary, the geostrophic relationship is obtained:

$$f^2 \eta - gH \left(\frac{\partial^2}{\partial x^2} + \frac{\partial^2}{\partial y^2} \right) \eta = \xi\,(x, y). \qquad (2.7)$$

In analogy with Equation (2.1), we find that $a = -gH$, $b = 0$ and $c = -gH$, which leads to $b^2 - 4ac = -4g^2 H^2 < 0$, and hence this PDE must be elliptic.

There are three different types of possible boundary conditions for this class of PDEs:

- u specified on the boundary (Dirichlet),
- $\partial u / \partial \vec{n}$, where $\vec{n}$ is the normal vector, specified on the boundary (Neumann),
- $au + \partial u / \partial \vec{n}$ specified on the boundary (Cauchy).

2.2 Parabolic equations

An example of a parabolic PDE is the diffusion equation for the dependent variable T corresponding to $e.g.$ temperature, salinity, humidity or any passive tracer:

$$\frac{\partial T}{\partial t} = K \frac{\partial^2 T}{\partial x^2} \quad ; \quad K > 0. \qquad (2.8)$$

Based on Equation (2.1), $a = K$, $b = 0$ and $c = 0$, leading to $b^2 - 4ac = 0$, and hence the PDE is parabolic.

Assume that $T\,(x, t)$ is the temperature distribution along the x-axis of a conducting rod as a function of time. To solve the equation over the interval $0 \le x \le L$ one must prescribe an initial condition $T\,(x, 0)$ on $0 \le x \le L$. The boundary conditions $T\,(0, t)$ and $T\,(L, t)$ must also be specified during the whole time period under consideration.

2.3 Hyperbolic equations

This class of PDEs describes wave motion. A typical example is the equation governing the vibrating string or the gravity waves in the ocean or atmosphere. In Section 2.1, we found that the shallow-water equations were elliptic in their steady state form. Here we will show that they can be hyperbolic in their time-dependent form. The wave equation can be derived from the simplest possible set-up of the shallow-water equations:

$$\frac{\partial u}{\partial t} = -g\frac{\partial \eta}{\partial x}, \tag{2.9a}$$

$$\frac{\partial \eta}{\partial t} = -H\frac{\partial u}{\partial x}. \tag{2.9b}$$

By eliminating u between the two equations we obtain the wave equation:

$$\frac{\partial^2 \eta}{\partial t^2} = gH\frac{\partial^2 \eta}{\partial x^2}. \tag{2.10}$$

In analogy with Equation (2.1) it is found that $b^2 - 4ac = 4gH > 0$ and hence this equation is a hyperbolic PDE.

The following first-order PDE, known as the advection equation, can also be classified as hyperbolic, since its solutions satisfy the wave equation:

$$\frac{\partial u}{\partial t} = -c_0\frac{\partial u}{\partial x}, \tag{2.11}$$

where $c_0 \equiv \sqrt{gH}$ is the speed of gravity waves. By first differentiating the advection equation with respect to t and then a second time with respect to x, it is possible to eliminate $\partial^2 u/\partial t\partial x$ between the resulting two equations and we obtain the wave equation.

In order to obtain unique solutions of Equation (2.10) for $0 \leq x \leq L$ we need boundary conditions at $x = 0$ and $x = L$, *viz.* the ends of the spatial domain, and two initial conditions. Possible boundary conditions are $u(0,t)$, $u(L,t)$, $\partial u(0,t)/\partial x$ or $\partial u(L,t)/\partial x$ specified at $x = 0$ and $x = L$. Necessary initial conditions are $u(x,0)$ and $\partial u(x,0)/\partial t$ specified over $0 \leq x \leq L$.

2.4 Overview

Having introduced these three types of PDEs, it is important to underline that the behaviour of the solutions, the proper initial and/or boundary conditions, and the numerical methods that can be used to find the solutions depend essentially on the type of PDE dealt with. Although non-linear multidimensional PDEs in general cannot be reduced to these canonical forms, we need to study these prototypes in order to develop an understanding of their properties, and then apply similar methods to the more complicated equations governing the motion of the ocean and atmosphere.

Exercises:

1. Solve the following equation by separation of variables:

$$\frac{\partial u}{\partial t} = -c\frac{\partial u}{\partial x}, \tag{2.12}$$

with the initial condition

$$u(x,0) = -Ae^{ikx}. \tag{2.13}$$

2. Show that the advection equation

$$\frac{\partial u}{\partial t} = -c\frac{\partial u}{\partial x} \tag{2.14}$$

has the general d'Alembert solution $u = f(x - ct)$, where f is an arbitrary continuously differentiable function. Interpret the equation geometrically in the xt-plane. Solve the equation with the following initial condition: $u(x,0) = g(x)$.

3. Derive the wave equation from the shallow-water equation system

$$\frac{\partial \vec{V}}{\partial t} = -g\nabla\eta, \tag{2.15a}$$

$$\frac{\partial \eta}{\partial t} = -H\nabla \cdot \vec{V}, \tag{2.15b}$$

where g is the gravitational acceleration, η the free-surface deviation from the equilibrium depth H, $\vec{V}$ the horizontal velocity vector (u, v) and $\nabla \equiv (\partial/\partial x, \partial/\partial y)$.

3. Finite Differences

3.1 The grid-point method

Let us study a function u with one independent variable x:

$$u = u(x).$$

Suppose we have an interval L, which is partitioned by $N + 1$ equally spaced grid points (including the two at the limits of the interval). The grid length is then $\Delta x = L/N$ and the grid points are located at $x_j = j\Delta x$, where $j = 0, 1, 2, ...N$ are integers. Let the value of u at x_j be represented by u_j.

3.2 Finite-difference schemes

The formal mathematical definition of the derivative of a function $u(x)$ is

$$\frac{du}{dx} = \lim_{\Delta x \to 0} \frac{u(x + \Delta x) - u(x)}{\Delta x}, \tag{3.1}$$

which is illustrated in Figure 3.1. A derivative and its finite-difference approximation differ in that in the latter case Δx remains finite and will not tend to zero. Note that in what follows, the term "finite difference" will be taken as synonymous with finite-difference approximations of derivatives.

We will now derive expressions which can be used to give an approximate value of a derivative at a grid point in terms of grid-point values. The finite-difference schemes can be constructed between values of u_j over the grid length Δx. As illustrated in Figure 3.2, the first derivative of $u(x)$ can be approximated in three ways:

How to cite this book chapter:
Döös, K., Lundberg, P., and Campino, A. A. 2022. *Basic Numerical Methods in Meteorology and Oceanography*, pp. 13–24. Stockholm: Stockholm University Press. DOI: https://doi.org/10.16993/bbs.c. License: CC BY 4.0

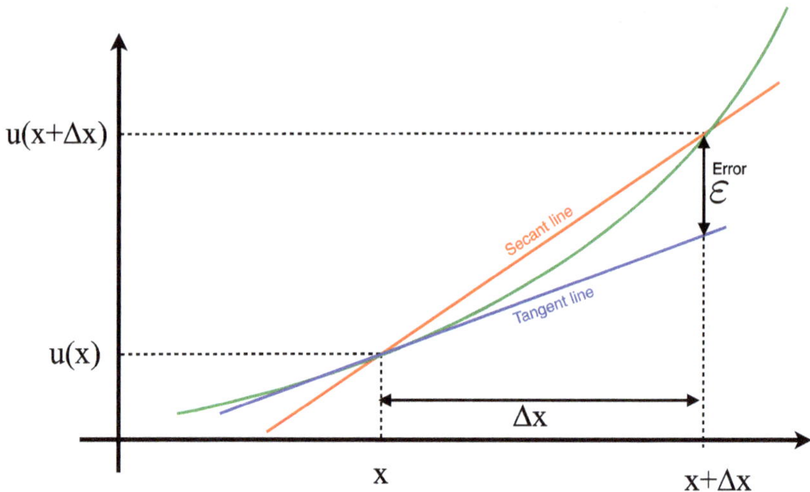

Figure 3.1. As Δx decreases, the red secant approaches the blue tangent, which is the derivative of the function u (green) at the point x when $\Delta x \to 0$. A finite difference is when $\Delta x \neq 0$.

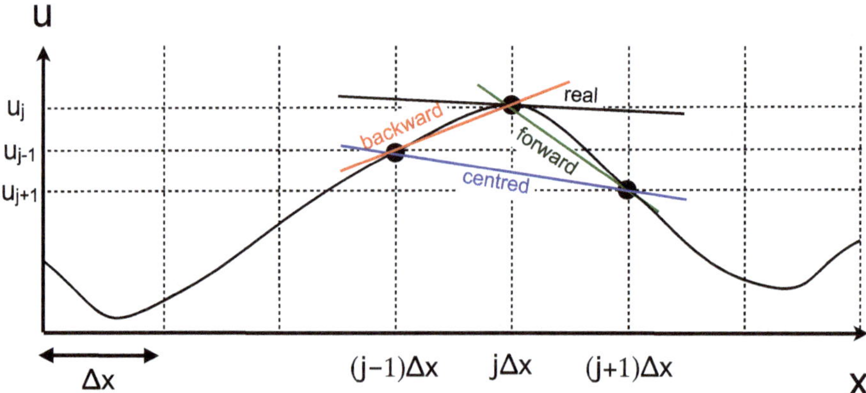

Figure 3.2. The Euler-backward, -forward and centred finite-difference approximations of the derivative of a function $u\left(x\right)$ at the point $j\Delta x$. The function is defined at the grid points $x = j\Delta x$ so that $u_j = u\left(x\right) = u\left(j\Delta x\right)$, where Δx is the grid length and $j = 0, 1, 2, \ldots$ are integers. Note that we have chosen an exceptionally difficult point around which to approximate the derivative, this to highlight potential difficulties with the method.

forward-difference scheme:

$$\left(\frac{du}{dx}\right)_j \approx \frac{u_{j+1} - u_j}{\Delta x},$$

centred-difference scheme:

$$\left(\frac{du}{dx}\right)_j \approx \frac{u_{j+1} - u_{j-1}}{2\Delta x},$$

backward-difference scheme:

$$\left(\frac{du}{dx}\right)_j \approx \frac{u_j - u_{j-1}}{\Delta x}.$$

These various schemes introduce errors that can estimated by deriving the finite-difference schemes in a more rigorous way using Taylor expansions. The Taylor series for $f(y)$ around $y = a$ is

$$f(y) = f(a) + (y-a) f'(a) + \frac{1}{2}(y-a)^2 f''(a) + \ldots + \frac{1}{n!}(y-a)^n f^{(n)}(a).$$
$$(3.2)$$

Note that a Taylor series is an expansion around a specific point, which yields a function that can be evaluated at another point.

3.2.1 Forward-difference scheme

Using the Taylor expansion given by Equation (3.2) and substituting $f(y)$ by $u(x)$, a by x_j and y by x_{j+1}, it is found that

$$u_{j+1} = u_j + \Delta x \left(\frac{du}{dx}\right)_j + \frac{1}{2}(\Delta x)^2 \left(\frac{d^2u}{dx^2}\right)_j + \frac{1}{6}(\Delta x)^3 \left(\frac{d^3u}{dx^3}\right)_j$$
$$+ \frac{1}{24}(\Delta x)^4 \left(\frac{d^4u}{dx^4}\right)_j + \frac{1}{120}(\Delta x)^5 \left(\frac{d^5u}{dx^5}\right)_j + \ldots . \quad (3.3)$$

The forward difference can now be expressed as

$$\frac{u_{j+1} - u_j}{\Delta x} = \left(\frac{du}{dx}\right)_j + \frac{1}{2}(\Delta x)\left(\frac{d^2u}{dx^2}\right)_j + \frac{1}{6}(\Delta x)^2\left(\frac{d^3u}{dx^3}\right)_j$$
$$+ \frac{1}{24}(\Delta x)^3\left(\frac{d^4u}{dx^4}\right)_j + \frac{1}{120}(\Delta x)^4\left(\frac{d^5u}{dx^5}\right)_j + \ldots . \quad (3.4)$$

The difference between this expression and the real derivative $(du/dx)_j$ is

$$\varepsilon = \frac{1}{2}(\Delta x)\left(\frac{d^2u}{dx^2}\right)_j + \frac{1}{6}(\Delta x)^2\left(\frac{d^3u}{dx^3}\right)_j$$
$$+ \frac{1}{24}(\Delta x)^3\left(\frac{d^4u}{dx^4}\right)_j + \frac{1}{120}(\Delta x)^4\left(\frac{d^5u}{dx^5}\right)_j + \ldots , \quad (3.5)$$

which is denoted the truncation error associated with the approximation of the derivative. The terms that have been truncated, *i.e.* "cut off", are represented by ε. Hence we have an accuracy of first order with

$$\varepsilon = \mathcal{O}\left(\Delta x\right), \tag{3.6}$$

which is the lowest order of accuracy that is acceptable. Note that we have assumed that Δx is small so that $(\Delta x)^m << (\Delta x)^n$ where m is greater than n. The higher-order terms in Δx can therefore be neglected.

3.2.2 Centred-difference scheme

The accuracy of the centred-difference scheme can be determined from Equation (3.3) and the analogous representation of u_{j-1}:

$$u_{j-1} = u_j - \Delta x \left(\frac{du}{dx}\right)_j + \frac{1}{2}\left(\Delta x\right)^2 \left(\frac{d^2u}{dx^2}\right)_j - \frac{1}{6}\left(\Delta x\right)^3 \left(\frac{d^3u}{dx^3}\right)_j$$
$$+ \frac{1}{24}\left(\Delta x\right)^4 \left(\frac{d^4u}{dx^4}\right)_j - \frac{1}{120}\left(\Delta x\right)^5 \left(\frac{d^5u}{dx^5}\right)_j + \dots, \tag{3.7}$$

so that

$$\frac{u_{j+1} - u_{j-1}}{2\Delta x} = \left(\frac{du}{dx}\right)_j + \frac{1}{6}\left(\Delta x\right)^2 \left(\frac{d^3u}{dx^3}\right)_j + \frac{1}{120}\left(\Delta x\right)^4 \left(\frac{d^5u}{dx^5}\right)_j + \dots.$$

Here the truncation error is of second order:

$$\varepsilon = \frac{1}{6}\left(\Delta x\right)^2 \left(\frac{d^3u}{dx^3}\right)_j + \dots = \mathcal{O}\left[\left(\Delta x\right)^2\right]. \tag{3.8}$$

3.2.3 Centred fourth-order difference scheme

A scheme with fourth-order accuracy can be obtained if we undertake the Taylor expansion given by Equation (3.2) and substitute $f(y)$ by $u(x)$, a by x_j and y by x_{j+2}:

$$u_{j+2} = u_j + 2\Delta x \left(\frac{du}{dx}\right)_j + \frac{1}{2}\left(2\Delta x\right)^2 \left(\frac{d^2u}{dx^2}\right)_j + \frac{1}{6}\left(2\Delta x\right)^3 \left(\frac{d^3u}{dx^3}\right)_j$$
$$+ \frac{1}{24}\left(2\Delta x\right)^4 \left(\frac{d^4u}{dx^4}\right)_j + \frac{1}{120}\left(2\Delta x\right)^5 \left(\frac{d^5u}{dx^5}\right)_j + \dots \tag{3.9}$$

and analogously,

$$u_{j-2} = u_j - 2\Delta x \left(\frac{du}{dx}\right)_j + \frac{1}{2}\left(2\Delta x\right)^2 \left(\frac{d^2u}{dx^2}\right)_j - \frac{1}{6}\left(2\Delta x\right)^3 \left(\frac{d^3u}{dx^3}\right)_j$$

$$+ \frac{1}{24}\left(2\Delta x\right)^4 \left(\frac{d^4u}{dx^4}\right)_j - \frac{1}{120}\left(2\Delta x\right)^5 \left(\frac{d^5u}{dx^5}\right)_j + \dots, \quad (3.10)$$

so that

$$\frac{u_{j+2} - u_{j-2}}{4\Delta x} = \left(\frac{du}{dx}\right)_j + \frac{4}{6}\left(\Delta x\right)^2 \left(\frac{d^3u}{dx^3}\right)_j + \frac{16}{120}\left(\Delta x\right)^4 \left(\frac{d^5u}{dx^5}\right)_j + \dots.$$

This scheme is, as the previous centred scheme, accurate to second order, and if we combine the two centred schemes so that

$$\frac{4}{3}\frac{u_{j+1} - u_{j-1}}{2\Delta x} - \frac{1}{3}\frac{u_{j+2} - u_{j-2}}{4\Delta x} = \left(\frac{du}{dx}\right)_j - \frac{1}{30}\left(\Delta x\right)^4 \left(\frac{d^5u}{dx^5}\right)_j + \dots,$$

$$(3.11)$$

we find an accuracy of fourth order, $i.e.$ $\varepsilon = \mathcal{O}\left[\left(\Delta x\right)^4\right]$.

3.2.4 Centred-difference scheme on a staggered grid

Finite differences on staggered grids (cf. Chapter 7) often require that the approximations be located between two contiguous grid points:

$$\left(\frac{du}{dx}\right)_{j+1/2} \approx \frac{u_{j+1} - u_j}{\Delta x}. \quad (3.12)$$

The accuracy of this scheme is obtained by first making a Taylor expansion around $x_{j+1/2}$ and then taking into account the point x_{j+1}, which leads to

$$u_{j+1} = u_{j+1/2} + \frac{\Delta x}{2}\left(\frac{du}{dx}\right)_{j+1/2} + \frac{1}{2}\left(\frac{\Delta x}{2}\right)^2 \left(\frac{d^2u}{dx^2}\right)_{j+1/2}$$

$$+ \frac{1}{6}\left(\frac{\Delta x}{2}\right)^3 \left(\frac{d^3u}{dx^3}\right)_{j+1/2} + \dots, \quad (3.13)$$

and then using the same Taylor expansion but with regard to the point x_j so that

$$u_j = u_{j+1/2} - \frac{\Delta x}{2} \left(\frac{du}{dx} \right)_{j+1/2}$$

$$+ \frac{1}{2} \left(\frac{\Delta x}{2} \right)^2 \left(\frac{d^2u}{dx^2} \right)_{j+1/2} - \frac{1}{6} \left(\frac{\Delta x}{2} \right)^3 \left(\frac{d^3u}{dx^3} \right)_{j+1/2} + \dots . \qquad (3.14)$$

Subtracting the latter equation from the former and dividing by Δx leads to

$$\frac{u_{j+1} - u_j}{\Delta x} = \left(\frac{du}{dx} \right)_{j+1/2} + \frac{1}{6} \left(\frac{\Delta x}{2} \right)^2 \left(\frac{d^3u}{dx^3} \right)_{j+1/2} + \dots . \qquad (3.15)$$

Hence the truncation error is of second order:

$$\varepsilon = \frac{1}{6} \left(\frac{\Delta x}{2} \right)^2 \left(\frac{d^3u}{dx^3} \right)_{j+1/2} + \dots = \mathcal{O} \left[(\Delta x)^2 \right]. \qquad (3.16)$$

Note that the only difference between this truncation error and the one for the previous centred finite difference given by Equation (3.8) is that this one is reduced, which is simply due to that the distance between the points is Δx instead of $2\Delta x$.

3.3 Time-difference schemes

The time-derivative schemes that are used for PDEs are often relatively simple, usually of second-order and sometimes even only of first-order accuracy. There are several reasons for this. First, it is a general experience that schemes constructed to have a high order of accuracy in time are mostly not very useful when solving PDEs. This is in contrast to the experience with ordinary differential equations, where very accurate methods, such as the Runge-Kutta scheme, are extremely successful. There is a basic reason for this discrepancy. With an ordinary differential equation of first order, the equation and a single initial condition is all that is required for an exact solution. Thus, the numerical-solution error is entirely due the degree of inadequacy of the scheme. With a PDE, the error associated with the numerical solution arises from both the shortcomings of the scheme and the insufficient information about the initial conditions, which only are known at discrete grid points. Thus, an increase in accuracy of the applied scheme improves only one of these two components, and the result is not too impressive.

Another reason for not using a finite-difference scheme of high accuracy of the time derivatives is that, in order to meet a stability requirement of the type to be discussed in the next chapter, it is usually necessary to choose a time step significantly smaller than that required for adequate accuracy. Once a time step has been specified, other errors, *e.g.* from the spatial differencing, are much greater that those due to the time differencing. Thus, computational effort is better spent in reducing these errors, and not in increasing the accuracy of the time-differencing schemes. This, of course, does not mean that it is not necessary to carefully consider the properties of various possible time-differencing schemes. Accuracy is only one important consideration when choosing a scheme.

To define some schemes, we consider a general first-order differential equation:

$$\frac{\partial u}{\partial t} = f(u, t), \qquad (3.17)$$

where typically $u = u(x, t)$. The independent variables x and t are space and time. f is thus a function of u, x and t, corresponding *e.g.* to the advection equation where $f = -c\, \partial u / \partial x$.

In order to discretise the equation we divide the time axis into segments of equal length Δt. The approximated value of $u(t)$ at time $t = n\Delta t$ is denoted u^n. In order to compute u^{n+1} we need to know at least u^n and often also u^{n-1}. A number of time-difference schemes are available.

3.3.1 Two-level schemes

These are schemes that use two different time levels, n and $n + 1$, so that the time integration yields

$$u^{n+1} = u^n + \int_{n\Delta t}^{(n+1)\Delta t} f(u, t)\, dt. \qquad (3.18)$$

The problem now is that f only exists as discrete values f^n and f^{n+1} at times $n\Delta t$ and $(n + 1)\Delta t$, respectively.

Euler-forward scheme

This is defined as

$$u^{n+1} = u^n + \Delta t\, f^n. \qquad (3.19)$$

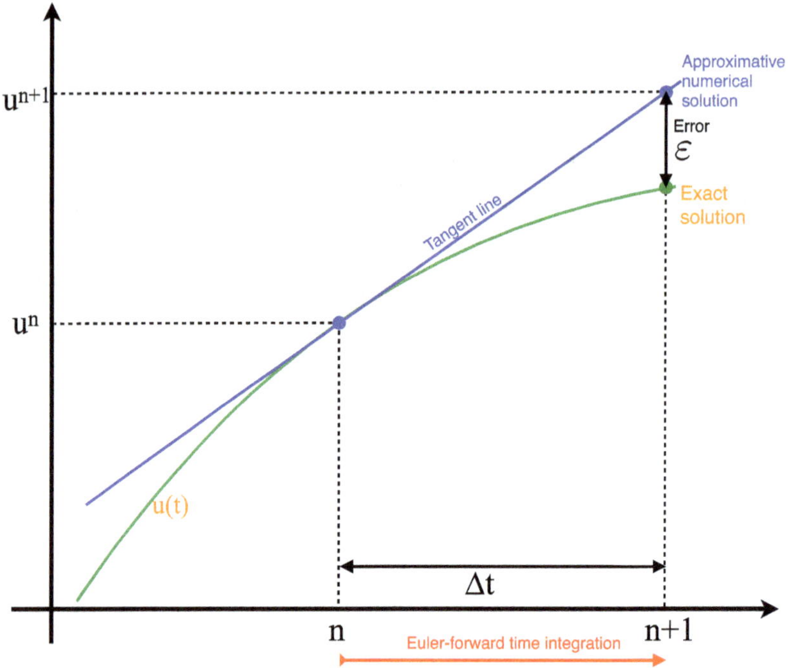

Figure 3.3. Time integration with the Euler-forward scheme.

Here the truncation error is $\mathcal{O}(\Delta t)$, *i.e.* the scheme is accurate to first order. It is said to be *uncentred*, since the "time derivative" pertains to time level $n + 1/2$ and the function to time level n. As in the previous section this Euler-forward scheme is of first-order accuracy (Figure 3.3).

Euler-backward scheme

This is defined as

$$u^{n+1} = u^n + \Delta t \, f^{n+1}. \tag{3.20}$$

The Euler-backward scheme is uncentred in time and is accurate to $\mathcal{O}(\Delta t)$. If, as here, a value of f is taken at time level $n+1$ and f depends on u, *i.e.* u^{n+1}, the scheme is said to be *implicit*. For an ordinary differential equation, it may be a straightforward matter to solve for u^{n+1}.

For a PDE it will, however, require solving a set of simultaneous equations, with one equation for each of the grid points of the computation region. If no value of f depends on u^{n+1} on the right-hand side of the equation above, the scheme is said to be *explicit*.

In the very simple cases when f only depends on t, such as e.g. $du/dt = -\gamma u$, the discretised equation becomes $u^{n+1} = u^n + \Delta t\left(-\gamma u^{n+1}\right)$, which can be rearranged as $u^{n+1} = u^n/(1+\gamma\Delta t)$ so that there are no terms at time level $n+1$ on the right-hand side. In this case the discretised equation can be integrated despite being implicit.

Crank-Nicolson scheme

The Crank-Nicolson scheme is based on the trapezoidal rule. If we approximate f by its average between time levels n and $n+1$ we obtain

$$u^{n+1} = u^n + \frac{1}{2}\Delta t\,(f^n + f^{n+1}). \tag{3.21}$$

This scheme is *implicit*, since it requires information from the future time level $n+1$. Note here that the finite-difference approximation is centred at $n+1/2$ between the two time steps n and $n+1$, *viz.*

$$\left(\frac{du}{dt}\right)^{n+1/2} = \frac{u^{n+1} - u^n}{\Delta t} + \varepsilon. \tag{3.22}$$

The truncation error ε can be found from the Taylor series in Equation (3.2). Substituting $f(y)$ by $u(t)$, a by $t^{n+1/2}$ and y by t^{n+1}, we obtain the following expression:

$$u^{n+1} = u^{n+1/2} + \frac{\Delta t}{2}\left(\frac{du}{dt}\right)^{n+1/2} + \frac{1}{2}\left(\frac{\Delta t}{2}\right)^2\left(\frac{d^2u}{dt^2}\right)^{n+1/2}$$
$$+ \frac{1}{6}\left(\frac{\Delta t}{2}\right)^3\left(\frac{d^3u}{dt^3}\right)^{n+1/2} + \dots. \tag{3.23}$$

Substituting $f(y)$ by $u(t)$, a by $t^{n+1/2}$ and y by t^n, we obtain the Taylor expansion of the function $u(t)$ at time level $n+1/2$:

$$u^n = u^{n+1/2} - \frac{\Delta t}{2}\left(\frac{du}{dt}\right)^{n+1/2}$$
$$+ \frac{1}{2}\left(\frac{\Delta t}{2}\right)^2\left(\frac{d^2u}{dt^2}\right)^{n+1/2} - \frac{1}{6}\left(\frac{\Delta t}{2}\right)^3\left(\frac{d^3u}{dt^3}\right)^{n+1/2} + \dots. \tag{3.24}$$

Subtracting Equation (3.24) from Equation (3.23) and dividing by Δt we obtain

$$\frac{u^{n+1} - u^n}{\Delta t} = \left(\frac{du}{dt}\right)^{n+1/2} + \frac{1}{24}(\Delta t)^2\left(\frac{d^3u}{dt^3}\right)^{n+1/2} + \dots, \tag{3.25}$$

which can also be expressed as :

$$\left(\frac{du}{dt}\right)^{n+1/2} = \frac{u^{n+1} - u^n}{\Delta t} + \varepsilon, \tag{3.26}$$

where

$$\varepsilon = -\frac{1}{24} (\Delta t)^2 \left(\frac{d^3 u}{dt^3}\right)^{n+1/2} + ... = \mathcal{O}\left[(\Delta t)^2\right]. \tag{3.27}$$

The truncation error ε is therefore of second order for the Crank-Nicolson scheme.

Matsuno's forward-backward scheme

To increase the accuracy compared to the Euler-forward and -backward schemes we can construct iterative schemes such as the Matsuno scheme, which is initiated by an Euler-forward time step:

$$u_*^{n+1} = u^n + \Delta t \, f^n. \tag{3.28}$$

In this case the value of the obtained u_*^{n+1} serves as an approximation of f^{n+1}, which hereafter is used to make a backward step to yield a final u^{n+1} :

$$u^{n+1} = u^n + \Delta t \, f_*^{n+1}, \tag{3.29}$$

where

$$f_*^{n+1} = f\left(u_*^{n+1}, (n+1)\Delta t\right). \tag{3.30}$$

This scheme is *explicit* and of accuracy $\mathcal{O}(\Delta t)$.

Heun scheme

This is similar to the Matsuno scheme and is explicit but of second-order accuracy, since the second step is made using the Crank-Nicolson scheme:

$$u_*^{n+1} = u^n + \Delta t \, f^n, \tag{3.31}$$

$$u^{n+1} = u^n + \frac{\Delta t}{2} (f^n + f_*^{n+1}). \tag{3.32}$$

The mid-point scheme

The mid-point scheme, also known as a second-order Runge-Kutta method, is like the Matsuno and Heun schemes a multi-stage scheme, which uses an intermediate estimate of the solution throughout the time-step. The scheme consists of first taking an Euler-forward scheme one half time step:

$$u^{n+1/2} = u^n + \frac{\Delta t}{2} f^n \qquad (3.33)$$

and then using the solution at the mid-point time level $n + 1/2$ to integrate with a centred scheme:

$$u^{n+1} = u^n + \Delta t\, f^{n+1/2}. \qquad (3.34)$$

The fourth-order Runge-Kutta scheme

The fourth-order Runge-Kutta scheme is similar to the previous scheme but integrates the solution with four steps instead of two. The scheme consists of first applying an Euler-forward scheme one half time step:

$$u_*^{n+1/2} = u^n + \frac{\Delta t}{2} f^n, \qquad (3.35)$$

and hereafter recomputing the solution at time level $n + 1/2$ with an Euler-backward scheme making use of $u_*^{n+1/2}$:

$$u^{n+1/2} = u^n + \frac{\Delta t}{2} f_*^{n+1/2}. \qquad (3.36)$$

Subsequently a centred scheme is used to arrive at time level $n + 1$ making use of $u^{n+1/2}$:

$$u_*^{n+1} = u^n + \Delta t\, f^{n+1/2}. \qquad (3.37)$$

The final solution is obtained by making use of Simpson's rule with the four values u^n, $u_*^{n+1/2}$, $u^{n+1/2}$ and u_*^{n+1}, resulting in

$$u^{n+1} = u^n + \frac{\Delta t}{6} \left(f^n + 2f_*^{n+1/2} + 2f^{n+1/2} + f_*^{n+1} \right). \qquad (3.38)$$

This fourth-order Runge-Kutta method is stable, and as its name indicates, accurate to fourth order. It is seldom used except in *e.g.* some regional high-resolution NWP models. Runge-Kutta methods of even higher order exist, but are rarely used in ocean or atmosphere circulation models.

3.3.2 Three-level schemes

These schemes use the time at three levels and the time integration becomes

$$u^{n+1} = u^{n-1} + \int_{(n-1)\Delta t}^{(n+1)\Delta t} f\left(u, t\right) dt. \qquad (3.39)$$

The simplest three-level scheme is to assign f a constant value equal to that at the middle of the time interval of length $2\Delta t$, which yields the leap-frog scheme

$$u^{n+1} = u^{n-1} + 2\Delta t\, f^n \tag{3.40}$$

of accuracy order $(\Delta t)^2$. This has been a widely used scheme in both atmospheric and oceanic models. Many of today's atmospheric circulation models use, however, Lagrangian time-stepping, which will be introduced in Chapter 11. Fourth-order schemes are, as mentioned above, sometimes also used. In some models such as the ROMS ocean model, there are several different time-schemes that can be used.

Exercise:

Determine the order of accuracy of the centred discretisations of the advection equation

$$\frac{u_j^{n+1} - u_j^{n-1}}{2\Delta t} + c\frac{u_{j+1}^n - u_{j-1}^n}{2\Delta x} = 0.$$

4. Numerical Stability

Here we will study partial differential equations (PDEs) with one dependent and two independent variables. Intuitively one often thinks of the advection equation as describing the evolution of a tracer drifting passively with the flow. However, we shall here consider various simplified forms of the advection equation describing the advection of a dependent variable. In practice, this has proved to be the most important part of the hydrodynamic equations for the atmosphere and the ocean.

We will use the advection equation to investigate what is required of the numerical schemes to yield stable solutions to the PDE, *i.e.* solutions such that small perturbations do not grow in time, but rather decrease.

4.1 The advection equation

This hyperbolic advection equation (cf. Section 2.3) is

$$\frac{\partial u}{\partial t} + c\frac{\partial u}{\partial x} = 0,$$

where $u = u(x, t)$ and c is the prescribed phase speed. Analytical solutions are of the form $u(x, t) = u_0 e^{ik(x-ct)}$, where $c \equiv \omega/k$, ω being the frequency and k the wave number.

Let us now consider one among many possible discretisations of this equation by using a centred difference scheme in both time and space:

$$\frac{u_j^{n+1} - u_j^{n-1}}{2\Delta t} + c\frac{u_{j+1}^n - u_{j-1}^n}{2\Delta x} = 0. \tag{4.1}$$

How to cite this book chapter:
Döös, K., Lundberg, P., and Campino, A. A. 2022. *Basic Numerical Methods in Meteorology and Oceanography*, pp. 25–39. Stockholm: Stockholm University Press. DOI: https://doi.org/10.16993/bbs.d. License: CC BY 4.0

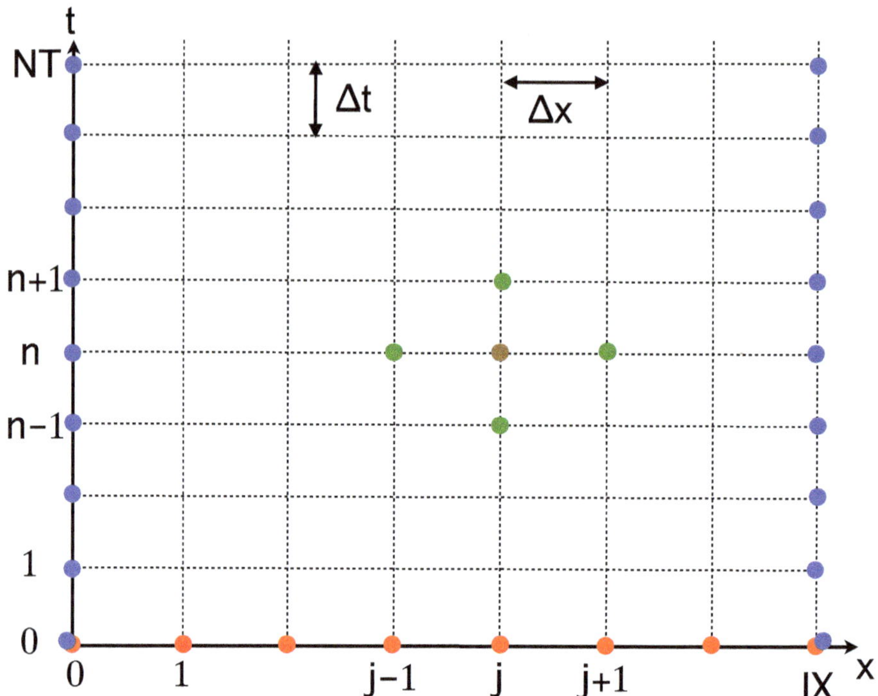

Figure 4.1. A finite-difference grid in time and space. The grid points used by the centred schemes in both time and space are shown as green dots. The red dots are the necessary initial-condition points and the blue dots illustrate the two boundaries, at which conditions must be prescribed. The brown dot shows the "here and now" point j, n.

Note that all the grid points for this scheme, which are shown by the green dots in Figure 4.1, are located around the "here and now" point j, n shown as the brown dot in the graph. When used in model applications, this needs to be reformulated in order to integrate values to the "future" time step $n + 1$:

$$u_j^{n+1} = u_j^{n-1} - \mu \left(u_{j+1}^n - u_{j-1}^n \right), \tag{4.2}$$

where

$$\mu \equiv \frac{c\Delta t}{\Delta x} \tag{4.3}$$

is known as the *Courant number*, sometimes denoted the CFL-number (Courant et al., 1928). Figure 4.2 shows how a cosine "hump" propagates, when integrated 25 time steps using Equation (4.2).

4.2 Initial and boundary conditions

We also need initial and boundary conditions in order to integrate Equation (4.2) forward in time on a grid such as that shown in Figure 4.1. The initial condition is that all the u_j^0 values (red dots) have to be prescribed, *i.e.* the variable u must have specified values at time step $n = 0$. Furthermore we need to be able to integrate a first time step, which is not possible with a three-level time scheme. We therefore use an Euler-forward scheme for the first time step and hereafter proceed using the leap-frog scheme for the rest of the time integration.

The boundary conditions imply that values for all the u_0^n and u_{JX}^n (blue dots in Figure 4.1) have to be prescribed, *i.e.* the variable u must have values at all time steps on the two boundaries located at $j = 0$ and $j = JX$, where JX is the total number of grid cells.

When we have a periodic domain, where *e.g.* the values on the eastern boundary equal those on the western boundary, which is the case for global models of the Earth, we can use periodic boundary conditions. For practical purposes this boundary can *e.g.* be located at the Greenwich meridian where the longitude can be expressed as $0°$ or $360°$. When u is computed in *e.g.* a Fortran code, at the eastern (j=0) and western (j=JX) boundaries one will need u-values for "j=-1" and "j=JX+1". This can easily be accomplished in the j-loops of the model code by introducing jm=j-1 and when jm=-1, replacing it by JX-1. The same procedure is used for the eastern boundary with jp=j+1 and when jp=JX+1, it is replaced by jp=1.

A segment of a Fortran code of the numerical time integration of the discretised advection equation (4.2) could look like this:

```fortran
mu=1. ! Courant number
u(:,0)=1. ! initialise the field to something, e.g. 1
! time loop
do n=1,NT-1
  do j=0,JX
    jp=j+1
    if(jp==JX+1) jp=1
    jm=j-1
    if(jm==-1) jm=JX-1
    u(j,n+1) = u(j,n-1) - mu * (u(jp,n)-u(jm,n))
  enddo
enddo
```

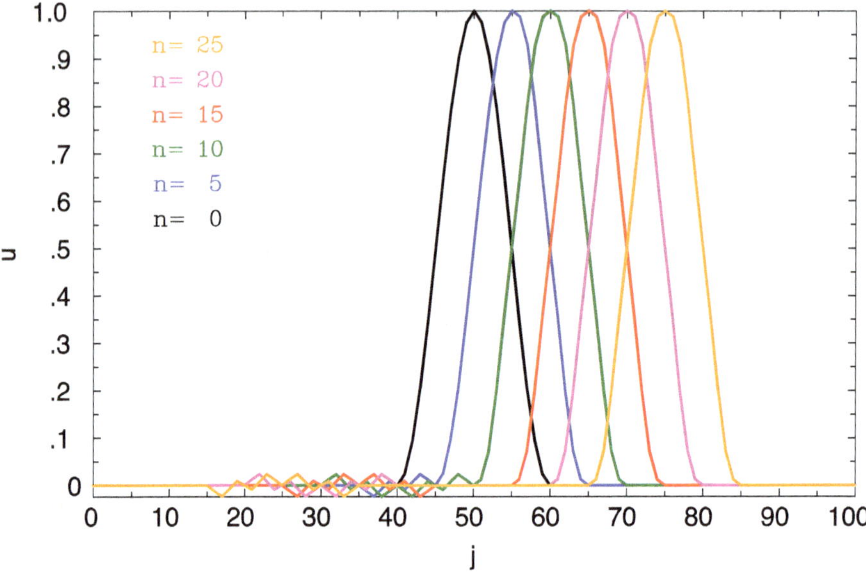

Figure 4.2. The advection equation integrated numerically with centred schemes in both time and space using Equation (4.1). The first time step has been integrated with an Euler-forward scheme. The initial value is $u_j^{n=0} = \cos\left[\pi(j - 50)/20\right]$ for $40 \leq j \leq 60$ and $u_j^{n=0} = 0$ for the remaining grid points. The Courant number for this scheme is $\mu \equiv c\Delta t/\Delta x = 1$. Solutions are graphed for time steps n=0, 5, 10, 15, 20, 25. Note the "numerical noise" propagating in the "wrong" direction in the form of small-amplitude jagged waves, a topic to be further considered in Chapter 6.

4.3 Stability analysis of the leap-frog scheme

In order to study the stability we will use what is known as the von Neumann method, which in fact was first briefly introduced by Crank and Nicolson (1947) and later more rigorously by Charney et al. (1950), in which latter study von Neumann actually was the last author. This method is generally not possible to use for non-linear equations and one is therefore limited to apply it to linearised versions of the equations in a numerical model. In general a solution of a linear equation can be expressed as a Fourier series, where each Fourier component is a solution. We can thus test the stability using solely one Fourier component of the form

$$u_j^n = u_0 e^{ik\left(j\Delta x - C_D n\Delta t\right)},$$

where k is the wave number. Note that the phase speed C_D, with the index D pertaining to finite differencing, is an approximation of the phase speed c occurring in the differential equation. It is this C_D, resulting from solving Equation (4.1), which here will be investigated. We note that

$$u_j^{n+1} = u_0 e^{ik\left[j\Delta x - C_D(n+1)\Delta t\right]} = u_j^n e^{-ikC_D\Delta t} = u_j^n \lambda,$$

where $\lambda \equiv e^{-ikC_D\Delta t}$ is the amplification factor. Similarly, for $n-1$ we find that

$$u_j^{n-1} = u_j^n \lambda^{-1}. \tag{4.4}$$

These results can be generalised as

$$u_j^{n+m} = u_j^n \lambda^m \tag{4.5}$$

and

$$u_j^n = \lambda^n u_0 e^{ikj\Delta x}. \tag{4.6}$$

From this we can deduce that if $|u|$ is not going to "blow up" when integrating in time, it is required that

$$|\lambda| = \left|e^{-ikC_D\Delta t}\right| \leq 1 \tag{4.7}$$

and reversely, if $|\lambda| > 1$ the solution is unstable and "blows up". For the stability condition to be fulfilled, C_D must be real. This technique for determining stability, based on examining the amplification factor λ, is known as the von Neumann method. We also need expressions for the Fourier components of the spatial derivatives:

$$u_{j+1}^n = u_0 e^{ik\left[(j+1)\Delta x - C_D n\Delta t\right]} = e^{ik\Delta x} u_j^n, \tag{4.8}$$

$$u_{j-1}^n = u_0 e^{ik\left[(j-1)\Delta x - C_D n\Delta t\right]} = e^{-ik\Delta x} u_j^n. \tag{4.9}$$

Let us now return to Equation (4.1) and introduce λ:

$$\frac{\lambda u_j^n - \lambda^{-1} u_j^n}{2\Delta t} + c\frac{e^{ik\Delta x} u_j^n - e^{-ik\Delta x} u_j^n}{2\Delta x} = 0, \tag{4.10}$$

which, since $e^{i\alpha} - e^{-i\alpha} = 2i\sin\alpha$, can be simplified to the quadratic equation

$$\lambda^2 + 2i\mu\sin(k\Delta x)\lambda - 1 = 0. \tag{4.11}$$

Furthermore $x^2 + \alpha x + \beta = 0$ yields that $x = -\alpha/2 \pm \sqrt{\alpha^2/4 - \beta}$, and hence Equation (4.11) has the solution

$$\lambda = -i\mu \sin(k\Delta x) \pm \sqrt{-(\mu \sin(k\Delta x))^2 + 1}. \qquad (4.12)$$

Keeping in mind that the absolute value of a complex number $a + ib$ is $\sqrt{a^2 + b^2}$, we find that if

$$[\mu \sin(k\Delta x)]^2 \leq 1, \qquad (4.13)$$

then

$$|\lambda|^2 = [\mu \sin(k\Delta x)]^2 + \left\{\sqrt{-[\mu \sin(k\Delta x)]^2 + 1}\right\}^2 = 1, \qquad (4.14)$$

i.e. the scheme we are presently examining is stable if Equation (4.13) holds true, a condition which also can be formulated as

$$|\mu \sin(k\Delta x)| \leq 1. \qquad (4.15)$$

Since $|\sin(k\Delta x)| \leq 1$, we have conditional stability if

$$|\mu| \leq 1 \; or \; |c| \leq \frac{\Delta x}{\Delta t}, \qquad (4.16)$$

this since we require stability for all wave numbers k and have hence chosen the "worst" case, *i.e.* when $\sin(k\Delta x) = 1$. The reverse case is when

$$[\mu \sin(k\Delta x)]^2 > 1. \qquad (4.17)$$

For simplicity we define

$$a \equiv \mu \sin(k\Delta x), \qquad (4.18)$$

so that Equation (4.17) is reduced to

$$a^2 > 1. \qquad (4.19)$$

It is then preferable to rewrite Equation (4.12) as

$$\lambda = -ia \pm i\sqrt{a^2 - 1} = i\left(-a \pm \sqrt{a^2 - 1}\right), \qquad (4.20)$$

or equivalently

$$|\lambda|^2 = \left(-a \pm \sqrt{a^2 - 1}\right)^2 = 2a^2 \pm 2a\sqrt{a^2 - 1} - 1, \qquad (4.21)$$

which has at least one root larger than one. Consequently when $\mu > 1$ $(c > \Delta x/\Delta t)$, then $|\lambda| > 1$ for at least some wave numbers k and thus the solution is unstable and hence of limited practical use.

The leap-frog scheme is thus *conditionally stable*. Given a spatial resolution Δx we require a time step Δt not exceeding $\Delta x/c$ for Equation (4.16) to be valid for the fastest possible phase speed in the system as illustrated by Figure 4.3. This is known as the *Courant-Friedrichs-Lewy* (CFL) criterion (Courant et al., 1928). In this particular case of the advection equation with centred finite differences both in time and space, the Courant number should not exceed 1 $(\mu \equiv c\Delta t/\Delta x \leq 1)$ in order for the CFL condition to be satisfied. This criterion will, as we shall see later, change depending on which equation and finite difference scheme we are dealing with.

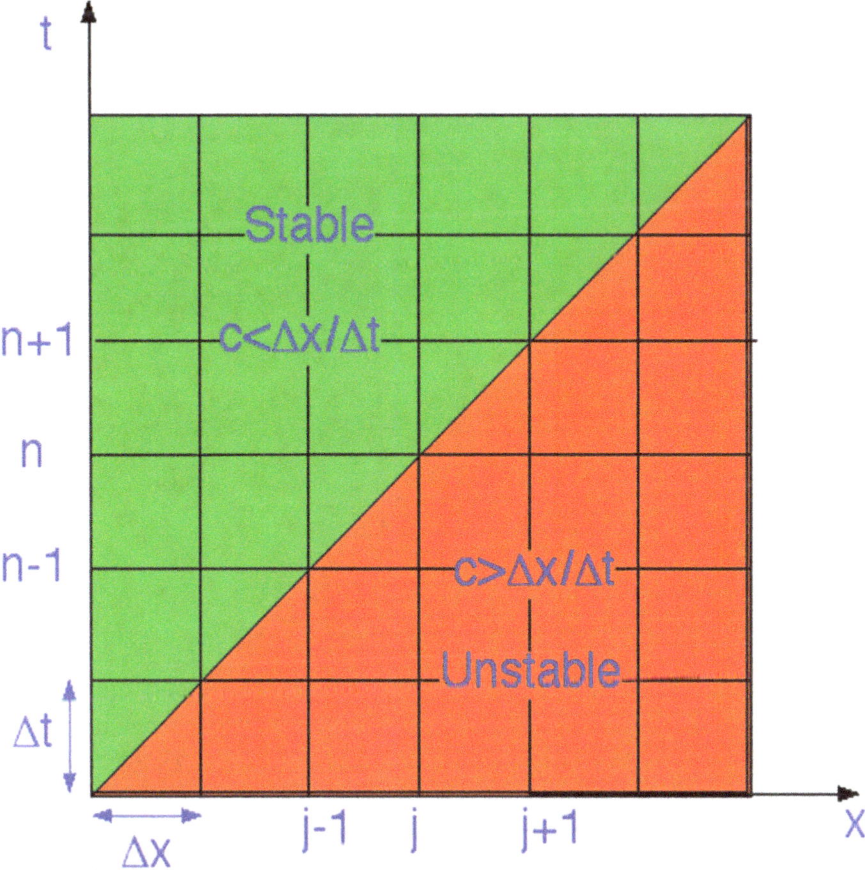

Figure 4.3. The Courant-Friedrichs-Lewy (CFL) stability criterion for schemes centred in both time and space.

An example of a successful integration of the advection equation with centred finite differences in space as well as in time has previously been given in Figure 4.2, where we see how the initially prescribed trigonometric "hump" with an amplitude equal to 1 progresses rightwards along the x-axis. This representation of a time-dependent process is, however, somewhat unwieldy and in practice what is known as a Hovmöller (1949) diagram is most frequently used, cf. Figure 4.4. This highly compact visualisation is based on graphing the the spatio-temporal evolution of the process in a coordinate system with the ordinate representing time and the abscissa pertaining to the spatial evolution of the process. The left-hand panel of Figure 4.4 thus shows precisely the same results as those given in Figure 4.2, where the gradually weakening coloration of the fringes of the "diagonal bands" do perfect justice to the spatial structure of the trigonometric "hump" in Figure 4.2. The right-hand panel of Figure 4.4 shows how for a Courant number larger than 1 instability sets in after less than 10 time steps and subsequently grows to encompass the entire spatial range under consideration.

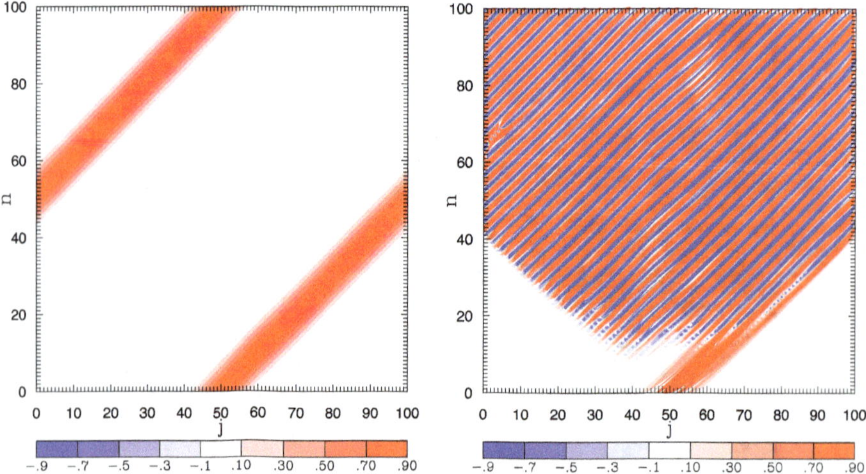

Figure 4.4. Hovmöller diagrams of the advection equation integrated numerically with a leap-frog scheme in time and a centred scheme in space, *viz.* Equation (4.1). The first time step has been integrated with an Euler-forward scheme. The initial value is $u_j^{n=0} = \cos\left[\pi(j-50)/20\right]$ for $40 \leq j \leq 60$ and $u_j^{n=0} = 0$ for the remaining grid points. The Courant number is $\mu \equiv c\Delta t/\Delta x = 1$ in the left panel, which yields a stable solution, but in the right panel the solution is seen to "blow up" for $\mu = 1.1$. Note that periodic boundary conditions have been applied.

Here it may finally be noted that if the advection equation dealt with here using centred finite differencing is applied to the ocean, then we, for a given spatial resolution Δx, must adjust the time step Δt to satisfy $\mu \leq 1$. Here the relevant phase speed is the one for long gravity waves $c = \sqrt{gH}$, where H is the depth of the ocean. In order to guarantee numerical stability we need to determine the maximum depth H_{MAX} and set the the time step to satisfy $\Delta t \leq \Delta x / c = \Delta x / \sqrt{gH_{MAX}}$.

4.4 Euler-forward scheme in time

Let us now apply an *ad hoc* numerical scheme to the advection equation, so that instead of Equation (4.1) we have

$$\frac{u_j^{n+1} - u_j^n}{\Delta t} + c\frac{u_{j+1}^n - u_{j-1}^n}{2\Delta x} = 0, \tag{4.22}$$

which yields

$$\lambda = 1 - i\mu \sin\left(k\Delta x\right). \tag{4.23}$$

As before the stability criterion is that $|\lambda| \leq 1$, and since the absolute value of λ satisfies

$$|\lambda|^2 = 1 + \left[\mu \sin\left(k\Delta x\right)\right]^2, \tag{4.24}$$

it is recognised that $|\lambda| > 1$. The solution consequently grows with time, independently of how we choose the time step. This scheme is thus *unconditionally unstable*, which is simply referred to as *unstable*.

4.5 The upstream scheme

This scheme is denoted upstream or upwind since it looks for information from where the wind or current comes by using two different spatial schemes depending on the direction of the velocity.

$$\frac{u_j^{n+1} - u_j^n}{\Delta t} + c\frac{u_j^n - u_{j-1}^n}{\Delta x} = 0 \quad if\ c > 0, \tag{4.25a}$$

$$\frac{u_j^{n+1} - u_j^n}{\Delta t} + c\frac{u_{j+1}^n - u_j^n}{\Delta x} = 0 \quad if\ c < 0. \tag{4.25b}$$

If $c > 0$ one should use the backward scheme in space from Equation (4.25a) in order to use information from where the flow comes. Reversely if $c < 0$ then one should use the forward scheme in space given by Equation (4.25b). Figures 4.6 and 4.7 show the time

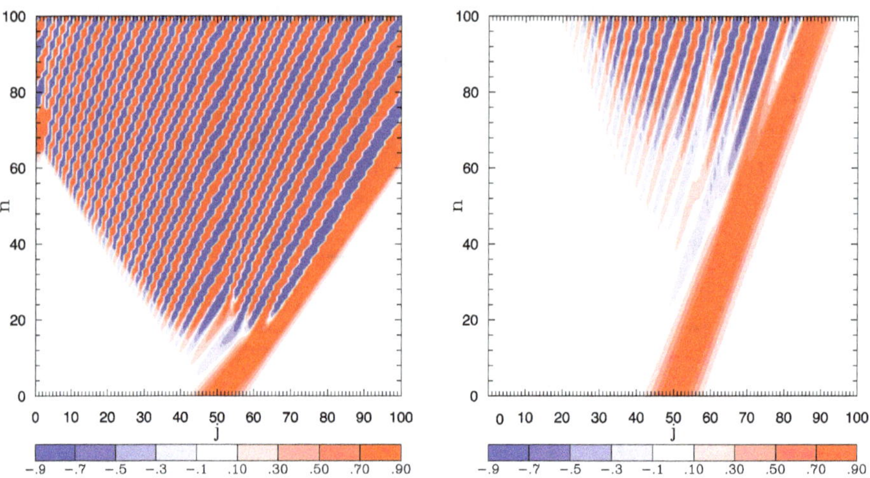

Figure 4.5. The advection equation integrated with an Euler-forward scheme in time and a centred scheme in space. In the left panel the Courant number is $\mu = 0.9$ and in the right panel 0.4 . Irrespective of the choice of Courant number, the solution will sooner or later "blow up".

evolution when Equation (4.25) is integrated with different Courant numbers μ.

A von Neumann stability analysis yields

$$\lambda = 1 - \mu \left[1 - \cos\left(k\Delta x\right) + i \sin\left(k\Delta x\right)\right],$$

resulting in

$$\begin{aligned}
|\lambda|^2 &= \left\{ 1 - \mu\left[1 - \cos\left(k\Delta x\right)\right]\right\}^2 + \left[\mu \sin\left(k\Delta x\right)\right]^2 \\
&= 1 + 2\mu\left[1 - \cos\left(k\Delta x\right)\right]\left(\mu - 1\right).
\end{aligned} \tag{4.26}$$

It is unfortunately not straightforward to see directly from this equation the corresponding stability criterion. For this we need to graph its amplification factor λ as a function of both resolution $k\Delta x$ and Courant number μ as in Figure 4.8. This shows that $|\lambda|^2 \leq 1$ for $0 \leq \mu \leq 1$, which also implies that c must be positive to ensure stability. The scheme is hence conditionally stable. This is similar to the leap-frog scheme but with the major difference that the amplification factor λ will be reduced when $\mu < 1$, which leads to a decrease of the amplitude at every new time step. This deamplification is clearly visible in the upper right panel of Figure 4.6 as well as in the right panel of Figure 4.7, where the upstream scheme has been integrated with $\mu = 0.5$. The

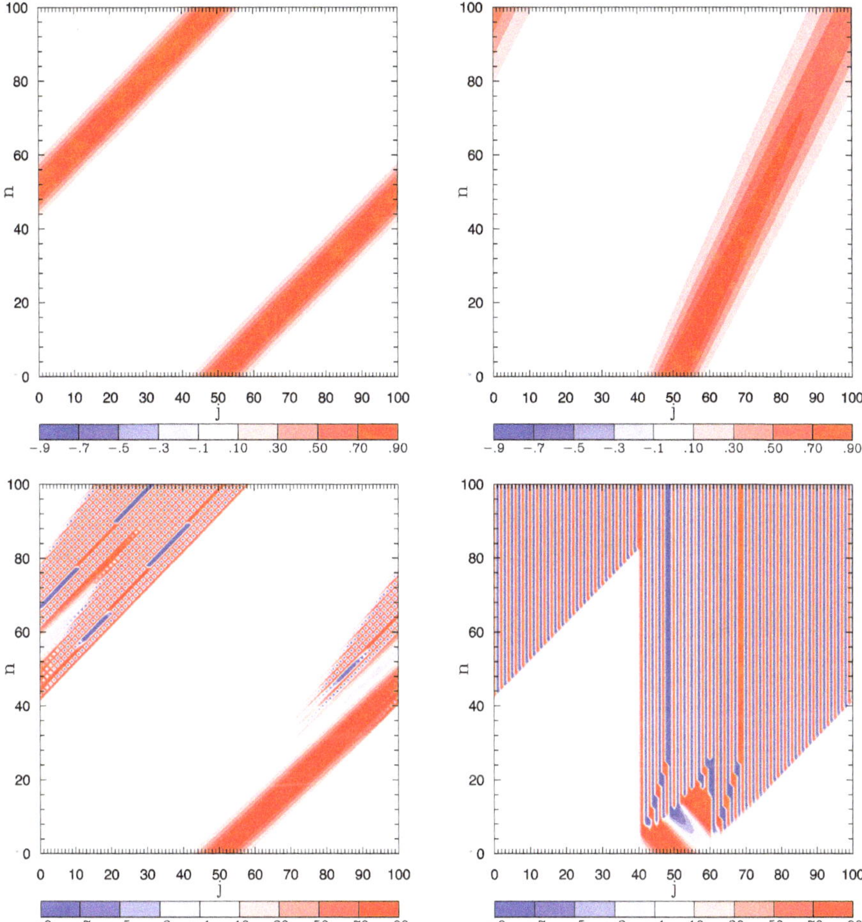

Figure 4.6. The advection equation integrated numerically with the upstream scheme given by Equation (4.25a). Stable solutions are shown in the upper panels, where to the left $\mu = 1$ and to the right $\mu = 0.5$, the latter solution being clearly dissipative. The lower panels show unstable solutions with to the left $\mu = 1.1$, and to the right $\mu = -1$, where in the latter case the spatial forward-difference scheme given by Equation (4.25b) should instead have been used.

decrease in amplitude is stronger for the short waves as illustrated by Figure 4.8. The upwind scheme should hence be integrated with a large Courant number μ, *i.c.* as close to one as possible. In this very simple case of the advection equation it is possible to use $\mu = 1$, but in a more comprehensive model, such as a GCM, this will not be possible and hence the scheme will lead to dissipation. The upwind

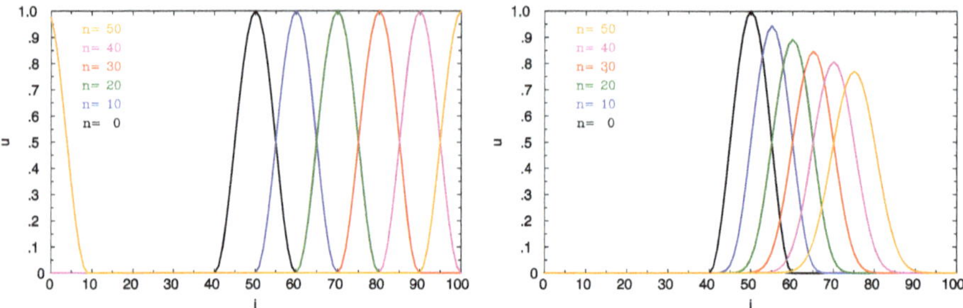

Figure 4.7. The amplitude u_j^n for different time steps n, computed with the upstream scheme given by Equation (4.25a), with $\mu = 1$ to the left and with $\mu = 0.5$ to the right. Note the rapid deamplification when integrated with $\mu = 0.5$.

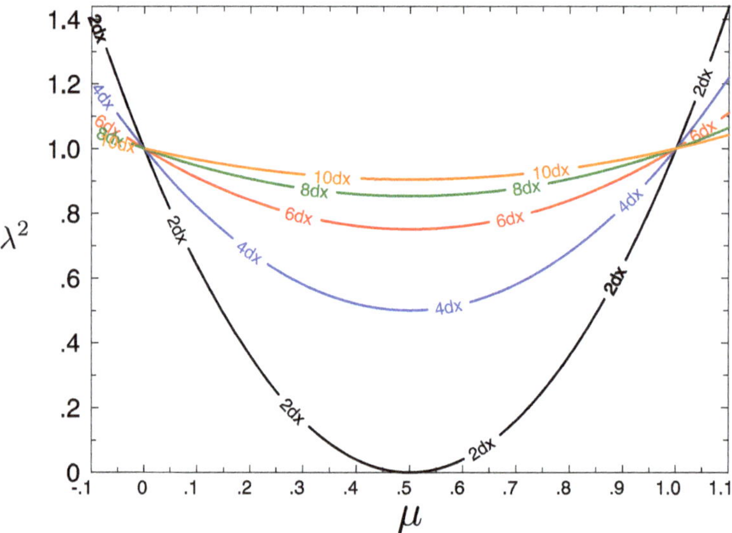

Figure 4.8. The squared amplification factor of Equation (4.26) as a function of the Courant number μ for different wavelengths measured in the number of grid lengths Δx. Note that there is a minimum at $\mu = 0.5$, which leads to a deamplification of, in particular, the short waves.

advection scheme was proposed by Courant et al. (1952) and used in early numerical weather-prediction models due to its good stability properties. It still finds use in idealised oceanic box models as well as in some general circulation models. The finite-volume model developed at ECMWF is *e.g.* using the upwind scheme in its advection-transport algorithm making use of an iterative approach (Smolarkiewicz et al., 2016). When using an upwind scheme one should, however, realise that it is highly diffusive. It uses backward spatial differencing if the velocity is in the positive x-direction, and forward spatial differencing for negative velocities. The term upwind denotes the use of upwind, or upstream, information when determining the form of the finite difference scheme; downstream information is ignored.

4.6 Stability analysis of the fourth-order scheme

Let us now study an advection scheme that uses the spatial scheme accurate to fourth order given by Equation (3.11):

$$\frac{u_j^{n+1} - u_j^{n-1}}{2\Delta t} + c\left(\frac{4}{3}\frac{u_{j+1}^n - u_{j-1}^n}{2\Delta x} - \frac{1}{3}\frac{u_{j+2}^n - u_{j-2}^n}{4\Delta x}\right) = 0. \qquad (4.27)$$

A von Neumann stability analysis yields

$$\lambda^2 + i\frac{\mu}{3}\left[8\sin\left(k\Delta x\right) - \sin\left(2k\Delta x\right)\right]\lambda - 1 = 0, \qquad (4.28\text{a})$$

which has the solution

$$\lambda = -\,i\frac{\mu}{6}\left[8\sin\left(k\Delta x\right) - \sin\left(2k\Delta x\right)\right]$$
$$\pm\sqrt{-\left[\frac{\mu}{6}\left[8\sin\left(k\Delta x\right) - \sin\left(2k\Delta x\right)\right]\right]^2 + 1}. \qquad (4.28\text{b})$$

If

$$\left[\frac{\mu}{6}\left[8\sin\left(k\Delta x\right) - \sin\left(2k\Delta x\right)\right]\right]^2 < 1, \qquad (4.29)$$

then

$$|\lambda|^2 = \left[\frac{\mu}{6}\left[8\sin\left(k\Delta x\right) - \sin\left(2k\Delta x\right)\right]\right]^2$$
$$+ \left\{\sqrt{-\left[\frac{\mu}{6}\left[8\sin\left(k\Delta x\right) - \sin\left(2k\Delta x\right)\right]\right]^2 + 1}\right\}^2 = 1, \qquad (4.30)$$

i.e. this scheme is stable if Equation (4.29) is fulfilled. This condition can also be expressed as

$$\left|\frac{\mu}{6}\left[8\sin\left(k\Delta x\right) - \sin\left(2k\Delta x\right)\right]\right| < 1$$

or

$$\mu < \frac{6}{8\sin\left(k\Delta x\right) - \sin\left(2k\Delta x\right)}. \qquad (4.31)$$

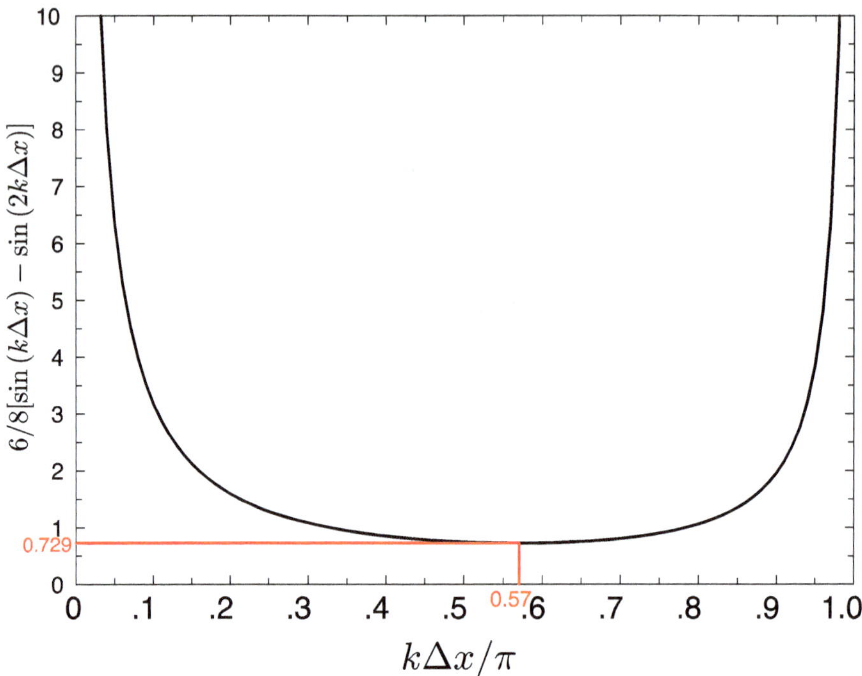

Figure 4.9. The right-hand side of Equation (4.31) as a function of $k\Delta x/\pi$. The minimum value in red, which is the stability criterion with a Courant number $\mu \equiv c\Delta t/\Delta x \approx 0.729$.

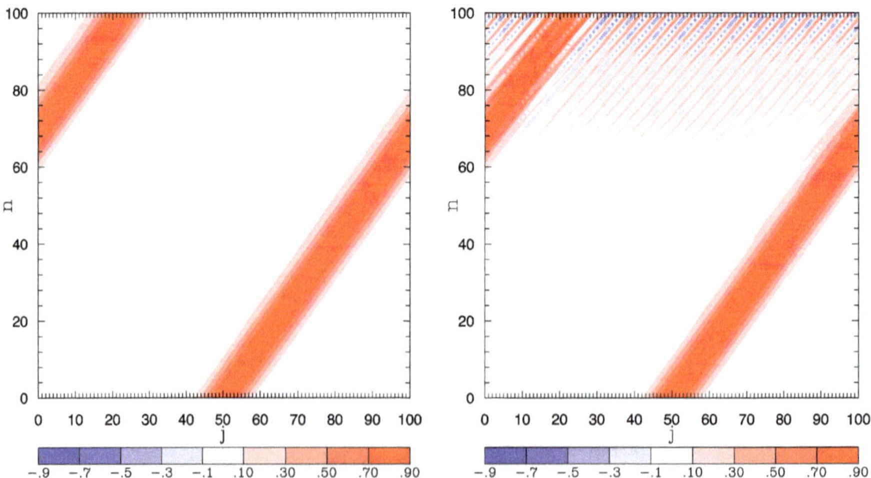

Figure 4.10. The advection equation integrated with the fourth-order spatial numerical scheme given by Equation (4.27). The solution is stable in the left panel where $\mu = 0.728$, but unstable in the right panel when $\mu = 0.730$.

The exact stability criterion is found by determining the $k\Delta x$ for which the right-hand side of Equation (4.31) attains its minimum as illustrated by Figure 4.9. This turns out to be for $k\Delta x/\pi \approx 0.57$, which leads to the scheme being stable for the Courant number $\mu \lesssim 0.729$. Figure 4.10 shows how the integration with the fourth-order scheme evolves in time with a Courant number just above and below its critical value.

Exercises:

1. Consider the leap-frog scheme for the advection equation:

$$\frac{u_j^{n+1} - u_j^{n-1}}{2\Delta t} + c\frac{u_{j+1}^n - u_{j-1}^n}{2\Delta x} = 0. \tag{4.32}$$

   Use

   $$u_j^n = \lambda^n u_0 e^{ikj\Delta x} \tag{4.33}$$

   and show that the amplification factor is

   $$\lambda = -i\mu \sin\left(k\Delta x\right) \pm \sqrt{1 - \left(\mu \sin\left(k\Delta x\right)\right)^2}. \tag{4.34}$$

2. Show that for $\mu > 1$ in the previous exercise, we will have one of the solutions to the differential equation "blowing up", at least for some wavelengths.

3. Discretise the advection equation with Euler-forward schemes in both time and space. Show that for $c > 0$ (backward scheme), the amplitude of the solutions will grow in time (these thus being unstable). But for $c < 0$ (forward scheme) the amplitude will decrease in time, $i.e.$ the solution is stable.

4. Undertake a stability analysis of the following discretisation of the advection equation:

$$\frac{u_j^{n+1} - \frac{1}{2}\left(u_{j+1}^n + u_{j-1}^n\right)}{\Delta t} + c\frac{u_{j+1}^n - u_{j-1}^n}{2\Delta x} = 0. \tag{4.35}$$

   Is it centred or uncentred?

5. The Computational Mode

Here we shall examine some of the consequences of partial differential equations having been approximated with finite-difference schemes.

5.1 The three-level scheme

One problem with a three-level scheme such as the leap-frog one is that it requires more than one initial condition to start the numerical integration. From a purely physical standpoint a single initial condition for $u^{n=0}$ should suffice. However, in addition to this physical initial condition, for computational purposes three-level schemes require an initial condition also for $u^{n=1}$ (in what follows the index j will be omitted since there is no spatial dependence in the equations to be dealt with). This value cannot be calculated using a three-level scheme, and will generally have to be determined using some type of two-level scheme. Consider the oscillation equation:

$$\frac{du}{dt} = i\omega u, \tag{5.1}$$

where $u = u(t)$. The analytical solution is

$$u = u_0 e^{i\omega t}. \tag{5.2}$$

Equation (5.1) can be integrated numerically with a leap-frog scheme:

$$u^{n+1} = u^{n-1} + 2i\omega \Delta t u^n. \tag{5.3}$$

If we now examine the amplification factor (cf. Chapter 4) we find

$$\lambda^2 - 2i\omega \Delta t \lambda - 1 = 0,$$

How to cite this book chapter:
Döös, K., Lundberg, P., and Campino, A. A. 2022. *Basic Numerical Methods in Meteorology and Oceanography*, pp. 41–48. Stockholm: Stockholm University Press. DOI: https://doi.org/10.16993/bbs.e. License: CC BY 4.0

which has the solution

$$\lambda_{1,2} = i\omega\Delta t \pm \sqrt{-(\omega\Delta t)^2 + 1}.$$

Thus there are two solutions of the form $u^{n+1} = \lambda u^n$. Since we are dealing with a linear equation, its solution will be a linear combination of these two:

$$u_1^n = \lambda_1^n u_1^0 \quad ; \quad u_2^n = \lambda_2^n u_2^0,$$

so that

$$u^n = a\lambda_1^n u_1^0 + b\lambda_2^n u_2^0, \tag{5.4}$$

where a and b are constants. If $u^{n+1} = \lambda u^n$ should represent the approximation of the true solution, then $\lambda \to 1$ when $\Delta t \to 0$. This condition is satisfied by λ_1, but also for $\lambda_2 \to -1$. The solution involving λ_1 is denoted the *physical mode* and that with λ_2 the *computational or numerical mode* induced by the numerical scheme. This latter mode changes sign for each even and odd n.

A straightforward way to illustrate the computational mode is to study the simple case when $\omega = 0$, *viz.*

$$\frac{du}{dt} = 0,$$

which has the exact solution

$$u(t) = constant.$$

Discretisation using the leap-frog scheme yields

$$u^{n+1} = u^{n-1}. \tag{5.5}$$

For a given physical initial condition $u^{n=0} = C_1$, we consider two special choices of $u^{n=1}$:

1. If calculating $u^{n=1}$ happens to yield the true value C_1, then for all n

    $$u^{n+1} = u^n \tag{5.6}$$

    or

    $$u^{n+1} = \lambda_1 u^n.$$

    In this case we have obtained a numerical solution identical to the true solution and which consists solely of the physical mode.

2. Suppose now instead that when calculating $u^{n=1}$ we obtain $u^{n=1} = -u^{n=0}$. Then for all n

$$u^{n+1} = -u^n \tag{5.7}$$

or

$$u^{n+1} = \lambda_2 u^n.$$

The numerical solution now consists entirely of the computational mode.

5.1.1 The computational initial condition

A suitable choice of the computational initial condition is of vital importance for obtaining a satisfactory numerical solution for short simulations, where the initial condition is crucial (as is the case for weather forecasts but less so for long climate simulations). The computational initial condition $(u^{n=1})$, which is one time step ahead of the physical condition $(u^{n=0})$, can be calculated with a single Euler-forward time step. Although this scheme is computationally unstable, it can be used for a single time step since a considerable number of steps are required before the solution finally "blows up".

The computational initial condition for our academic oscillation-equation case is then, with an Euler-forward time step:

$$u^{n=1} = u^{n=0} + i\omega\Delta t u^{n=0}. \tag{5.8}$$

An alternative computational initial condition, useful when the solution is less sensitive to the initial condition, is to assign the same value to both time steps $(u^{n=1} = u^{n=0})$, but, as shown above, this can immediately trigger a computational mode.

5.2 Suppression of the computational mode

The computational mode induced by the leap-frog scheme can be suppressed in two different ways:

5.2.1 Euler-forward or -backward schemes at regular intervals

The computational-mode problem can be solved by integrating with an Euler-forward or -backward scheme at regular intervals constituted by a certain number of time steps (50 or so are often used). For Equation (5.5) from the previous example this would imply that $u^{n+1} = u^n$ is

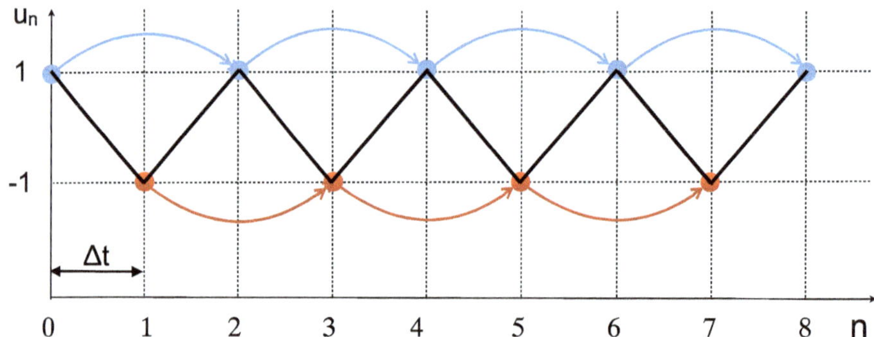

Figure 5.1. Illustration of the physical mode in blue from Equation (5.6) and the computational mode in red from Equation (5.7). The initial physical condition is set to $u^{n=0} = 1$ and the computational initial condition is set to $u^{n=1} = -1$. The resulting solution in black will thus change sign every time step since the two modes are uncoupled.

taken every 50 time steps, which eliminates the computational mode so that in Equation (5.4) $a = 1$ and $b = 0$.

5.2.2 The Robert-Asselin filter

Another way of suppressing the computational mode, which is the most common one used in atmospheric models, is to employ a Robert-Asselin filter (Robert, 1966; Asselin, 1972). First one applies a leap-frog integration to

$$\frac{\partial u}{\partial t} = g\left(u\right), \tag{5.9}$$

this in order to obtain the solution at time level $n + 1$:

$$u^{n+1} = u_f^{n-1} + 2\Delta t\, g\left(u^n\right), \tag{5.10}$$

whereafter the filter is applied as a smoothing between the three time levels $n - 1$, n and $n + 1$ so that

$$u_f^n = u^n + \gamma\left(u_f^{n-1} - 2u^n + u^{n+1}\right), \tag{5.11}$$

where the index f indicates the filtered values and γ is the Asselin coefficient, usually chosen to range between 0.01 and 0.2. The next "frog jump" will be

$$u^{n+2} = u_f^n + 2\Delta t\, g\left(u^{n+1}\right). \tag{5.12}$$

A segment of a Fortran code of the numerical time integration of the discretised advection equation (4.2) employing a Robert-Asselin filter could look like this:

```fortran
gamma = 0.01 ! Robert-Asselin filter coefficient
mu = 1. - gamma ! Courant number
! time loop
do n=1,NT-1
 ! Leap-frog scheme of the advection equation
 do i=1,JX-1
  u(i,n+1) = u(i,n-1) - mu * (u(i+1,n)-u(i-1,n))
 enddo
! Robert-Asselin filter
 do i=1,JX-1
  u(i,n) = u(i,n)+gamma*(u(i,n-1)-2.*u(i,n)+u(i,n+1))
 enddo
enddo
```

Note that the added term resembles smoothing in time; an approximation of an ideally time-centred smoother is

$$u_f^n = u^n + \gamma \left(u^{n-1} - 2u^n + u^{n+1} \right). \tag{5.13}$$

In our particular case of the discretised oscillation given by Equation (5.3) we can estimate the damping effect of the Robert-Asselin filter by introducing the discretised solution $u^n = u_0 e^{i\omega n \Delta t}$ into the smoother (Equation (5.13)), with the exception that u^{n-1} is taken as an unfiltered value. This results in

$$u_f^n = u^n \left[1 - 4\gamma \sin^2 \left(\omega \Delta t / 2 \right) \right]. \tag{5.14}$$

The computational mode, the period of which is $2\Delta t$, is hence reduced by $(1 - 4\gamma)$ every time step. Since the field at $n - 1$ is replaced by an already filtered field, the Robert-Asselin filter introduces a slight difference compared to this simplified filter.

Another drawback of the Robert-Asselin filter is that it affects the stability of the schemes. A stability analysis can be performed in the case of the advection equation discretised with centred schemes (same as Equation 4.2):

$$u_j^{n+1} = u_{j\,f}^{n-1} - \mu \left(u_{j+1}^n - u_{j-1}^n \right), \tag{5.15}$$

with the filter of Equation (5.11). The amplification factor then becomes

$$\lambda = -ia + \gamma \pm \sqrt{(1 - \gamma)^2 - a^2}, \tag{5.16}$$

where $a \equiv \mu \sin(k\Delta x)$. For stability we require as usual that $|\lambda| \leq 0$, which is obtained when $\mu \leq 1-\gamma$ or $\mu+\gamma \leq 1$. The adding of a Robert-Asselin filter results hence in a stricter stability condition and requires a shorter time step for a given spatial resolution than without filter.

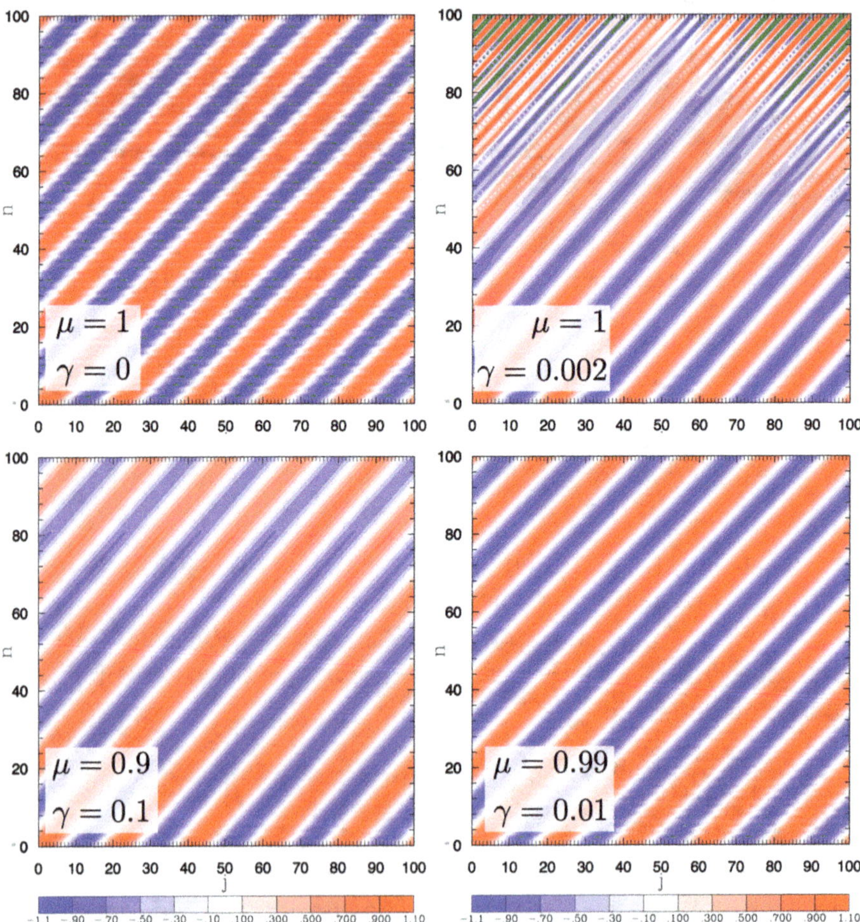

Figure 5.2. Hovmöller diagrams of the advection equation, integrated with the leap-frog scheme and a Robert-Asselin filter. The solution at the first time step has been set to equal the initial condition instead of integrating with an Euler-forward in order to generate a computational mode. Top left panel with no filter and an oscillating solution due to the computational mode. Top right panel with an unstable condition ($\mu + \gamma \leq 1$). Lower left panel with too much filtering but stable. Lower right panel with just enough Robert-Asselin filter to smooth the solution and a Courant number to match γ for stability.

In Figure 5.2, we have illustrated the Robert-Asselin filter for 4 different cases, where the advection equation has been integrated with the leap-frog scheme of Equation (5.15). The initial condition is a cosine wave. The first time step has not been integrated as usual with an Euler-forward scheme but set to be constant in time. This in order to immediately generate a computational mode, which we then try to filter out. The first case is with no filter and a persistent computational mode. The second case is with a filter but without respecting the stricter stability criterion, which makes the solution unstable and the wave "blows up". The third case is with a too strong filter, which dampens the wave too much although it filters out the computational mode. The fourth case is with just enough filter to eliminate the computational mode without too much damping.

Another test case is illustrated in Figure 5.3, where a shallow-water model has been integrated with different coefficients of the Robert-Asselin filter.

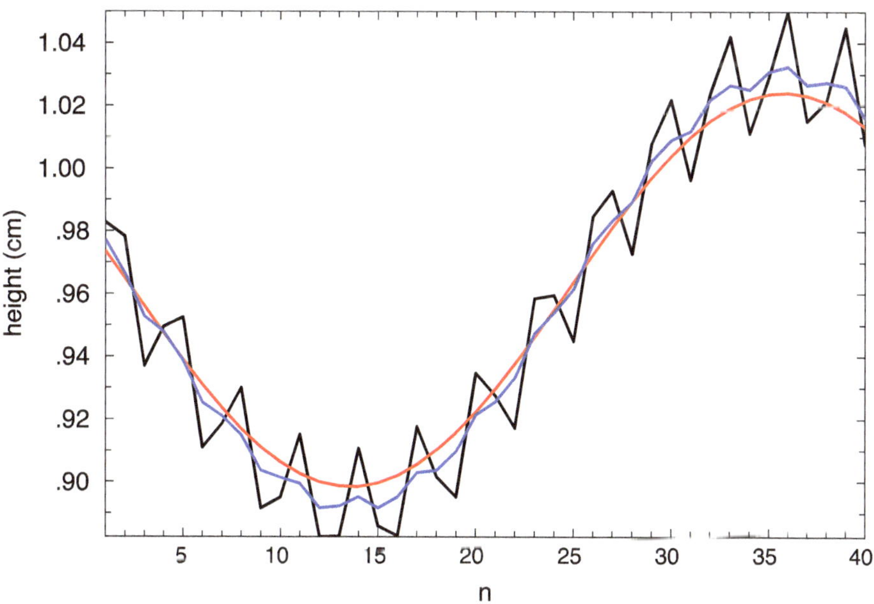

Figure 5.3. Time evolution of a shallow-water model integrated without a Robert-Asselin filter in black, with a Robert-Asselin filter $\gamma = 0.1$ in red and with $\gamma = 0.01$ in blue. Note that the computational mode vanishes completely when the stronger filter is applied, but the amplitude of the solution is at the same time damped.

5.2.3 The Robert-Asselin-Williams filter

It was recognised that when used with the leap-frog scheme, the feature of the Robert-Asselin filter of not conserving the mean degrades the numerical accuracy. Williams (2009) tackled this problem by introducing an extra step in the filtering process in order to include the possibility of conserving the mean value. The resulting filter is implemented in leap-frog integrations as follows:

$$u^{n+1} = u_{ff}^{n-1} + 2\Delta t\, g\left(u_f^n\right), \tag{5.17}$$

$$u_{ff}^n = u_f^n + \frac{\gamma\alpha}{2}\left(u^{n+1} - 2u_f^n + u_{ff}^{n-1}\right), \tag{5.18}$$

$$u_f^{n+1} = u^{n+1} - \frac{\gamma(1-\alpha)}{2}\left(u^{n+1} - 2u_f^n + u_{ff}^{n-1}\right). \tag{5.19}$$

This Robert-Asselin-Williams filter introduces an extra operation that is straightforward to implement and does not represent a significant computational expense compared to the Robert-Asselin filter. It also introduces a new parameter, α, such that $0 < \alpha < 1$, where $\alpha = 1$ corresponds to the traditional Robert-Asselin filter. Williams (2009) showed that a value of $\alpha = 0.53$ minimises spurious numerical impacts on the physical solution and yields the closest match to the exact solution of the equation under consideration over a broad frequency range.

6. The Computational Phase Speed

We shall now investigate the accuracy of the computational phase speed associated with using discretisations in space as well as in time.

6.1 Dispersion due to the spatial discretisation

Let us first examine the advection equation with a centred scheme in space:

$$\frac{\partial u}{\partial t} + c\frac{u_{j+1}^n - u_{j-1}^n}{2\Delta x} = 0.$$

By inserting a wave solution

$$u_j(t) = u_0 e^{ik\left(j\Delta x - C_D t\right)},$$

we find

$$C_D = c\frac{\sin(k\Delta x)}{k\Delta x},$$

where C_D is the computational phase speed and c the prescribed analytical phase speed. Their ratio should ideally be as close as possible to one, but is

$$\frac{C_D}{c} = \frac{\sin(k\Delta x)}{k\Delta x}.$$

The computational group velocity is

$$C_{Dg} = \frac{d(\omega_D)}{dk} = \frac{d(kC_D)}{dk} = c\cos(k\Delta x),$$

which is dispersive since it depends on the wave number k, cf. Figure 6.1.

How to cite this book chapter:

Döös, K., Lundberg, P., and Campino, A. A. 2022. *Basic Numerical Methods in Meteorology and Oceanography*, pp. 49–55. Stockholm: Stockholm University Press. DOI: https://doi.org/10.16993/bbs.f. License: CC BY 4.0

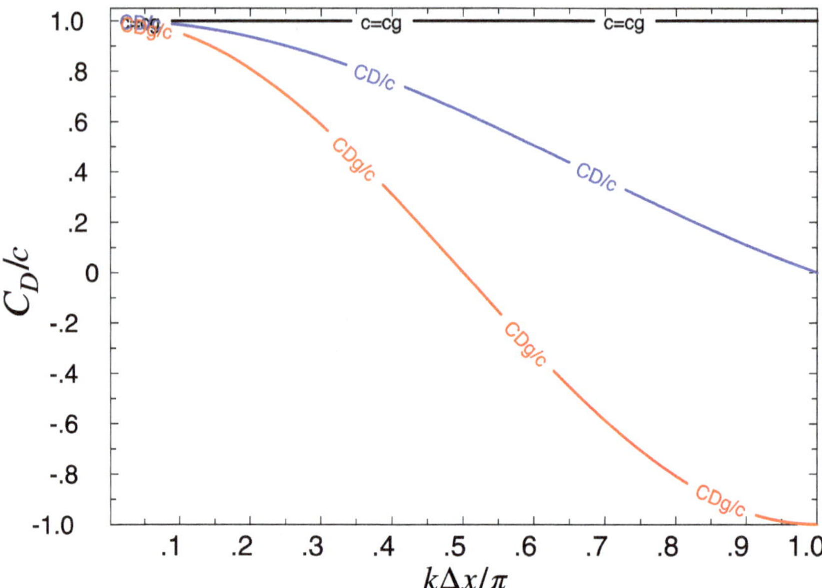

Figure 6.1. The computational phase speed associated with centred finite differencing in space compared to the analytical phase speed. The black line represnts the solution of the continuous equations which pertains the nondispersive analytical case, *i.e.* the phase speed is the same as the group velocity $c = c_g$. The blue curve is the computational phase speed normalised by dividing with c. Note that when the wave number increases (*i.e.* the wavelength decreases) the computational phase speed deviates from the analytical phase speed. The phase speed is clearly dispersive since the waves propagate at different speeds depending on their wavelengths. The red curve shows the analogously normalised computational group velocity C_{Dg} which at wavelengths shorter than 4 grid cells ($k\Delta x < \pi/2$) is in the opposite direction of c_g.

6.2 Dispersion due to the time discretisation

The effects of the centred scheme in time will be analysed in the same way as done for the effects of the centred scheme in space:

$$\frac{u_j^{n+1} - u_j^{n-1}}{2\Delta t} + c\frac{\partial u}{\partial x} = 0.$$

Inserting the wave solutions

$$u^n (x) = u_0 e^{ik(x - C_D n\Delta t)},$$

we obtain

$$C_D = \frac{\arcsin(\omega \Delta t)}{k \Delta t},$$

where C_D is the computational phase speed and c the analytical phase speed. Their ratio should ideally be as close as possible to one, but is

$$\frac{C_D}{c} = \frac{\arcsin(\omega \Delta t)}{\omega \Delta t}.$$

The computational and analytical group velocities also differ, so that

$$\frac{C_{Dg}}{c_g} = \frac{d(\omega_D)}{c_g dk} = \frac{d(kC_D)}{c_g dk} = \frac{1}{\sqrt{-(\omega \Delta t)^2 + 1}}.$$

The computational phase speed and group velocity (both normalised with c) are presented in Figure 6.2, graphed as functions of $\omega \Delta t$.

6.3 Dispersion due to spatial and temporal resolution

Let us now investigate the effects of the leap-frog scheme on the phase speed and group velocities using the advection equation with centred schemes in both time and space:

$$\frac{u_j^{n+1} - u_j^{n-1}}{2 \Delta t} + c \frac{u_{j+1}^n - u_{j-1}^n}{2 \Delta x} = 0.$$

By inserting a wave solution

$$u_j^n = u_0 e^{ik\left(j \Delta x - C_D n \Delta t\right)},$$

we obtain

$$C_D = \frac{1}{k \Delta t} \arcsin\left[\frac{c \Delta t}{\Delta x} \sin(k \Delta x)\right],$$

where C_D is the computational phase speed and c the analytical phase speed. Their ratio should ideally be as close as possible to one, but is

$$\frac{C_D}{c} = \frac{1}{\mu k \Delta x} \arcsin\left[\mu \sin(k \Delta x)\right], \tag{6.1}$$

where the Courant number is, as previously defined, $\mu \equiv c \Delta t / \Delta x$.

This computational phase speed C_D is a function of the wave number k and the resolution Δx. The finite differencing in space thus causes a *computational dispersion*. As $k \Delta x$ increases, C_D decreases from c to

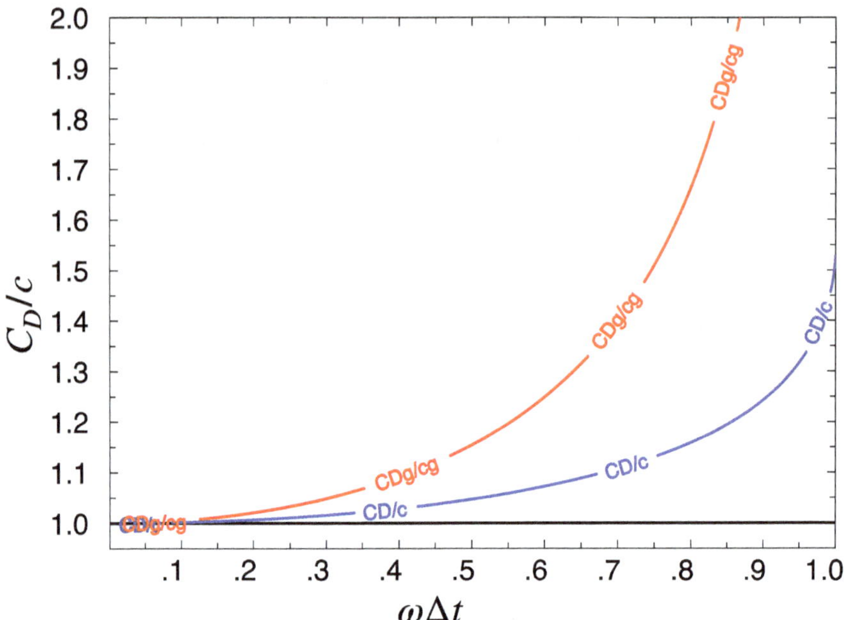

Figure 6.2. The computational phase speed associated with centred finite differencing in time compared to the analytical phase speed. The black line represents the solution of the continuous equations which is the nondispersive analytical case, *i.e.* the phase speed is the same as the group velocity, *viz.* $c = c_g$. The blue curve is the computational phase speed normalised by dividing with c. Note that when the time step increases relative to the wave frequency ω, the computational phase speed increases and deviates from the analytical phase speed. The red curve shows the computational group velocity C_{Dg}.

zero when $k\Delta x = \pi$, which corresponds to the shortest possible wave with a wavelength of two grid cells ($\lambda = 2\Delta x$). Thus, all waves propagate at a slower speed than the analytical phase speed c, with this decelerating effect increasing as the wavelength decreases. The two-grid-cell wave is stationary. Note that if $\mu = 1$, which is the Courant-number limit of stability for the advection equation, the computational phase speed is the same as the analytical one, *viz.* $C_D = c$.

The reason for the two-grid-cell wave being stationary is obvious when looking at the wave illustrated in Figure 6.3. For this wave we have $u_{j+1} = u_{j-1}$ at all grid points, corresponding to $\partial u_j/\partial t = 0$ in the advection equation.

The computational group velocity is here

$$C_{Dg} = \frac{d\omega_D}{dk} = \frac{d\left(kC_D\right)}{dk} = \frac{d}{dk}\left\{\frac{1}{\Delta t}\arcsin\left[\mu\sin\left(k\Delta x\right)\right]\right\}. \qquad (6.2)$$

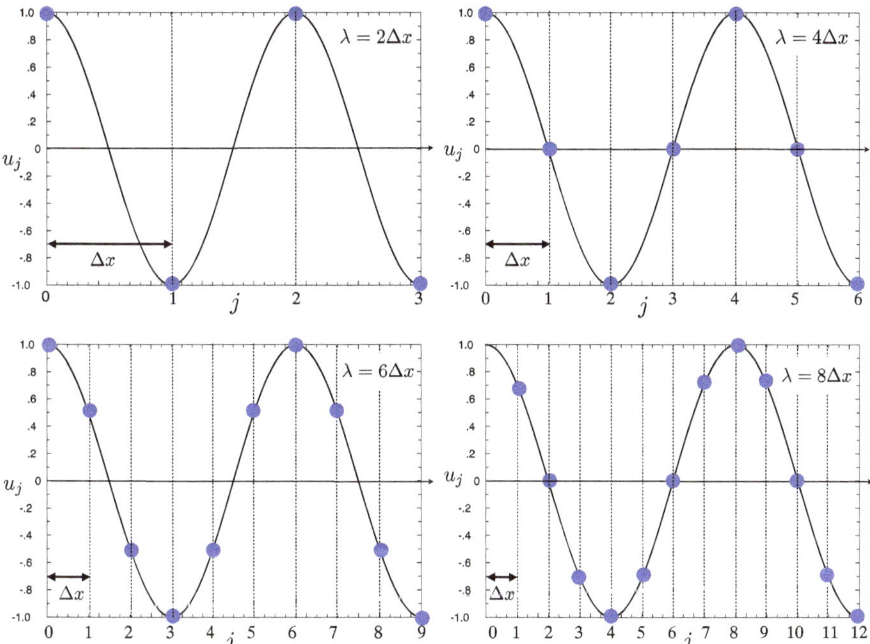

Figure 6.3. The two, four, six and eight grid-interval wave with a wavelength of $\lambda = 2, 4, 6$ and $8\Delta x$.

Noting that the derivative of the inverse sine function is

$$\frac{d}{dx}\arcsin\left[f\left(x\right)\right] = \frac{1}{\sqrt{1-f^2}}\frac{df}{dx}, \qquad (6.3)$$

we obtain

$$C_{Dg} = \frac{1}{\Delta t}\left(\frac{1}{\sqrt{1-\left[\mu\sin\left(k\Delta x\right)\right]^2}}\right)\frac{d}{dk}\left[\mu\sin\left(k\Delta x\right)\right]$$

$$= \frac{c\cos\left(k\Delta x\right)}{\sqrt{1-\left[\mu\sin\left(k\Delta x\right)\right]^2}}. \qquad (6.4)$$

Both computational speeds are functions of the wave number, and thus we recognise that the spatial differencing again results in computational dispersion. Since the analytical group velocity is $c_g = c$ it makes sense

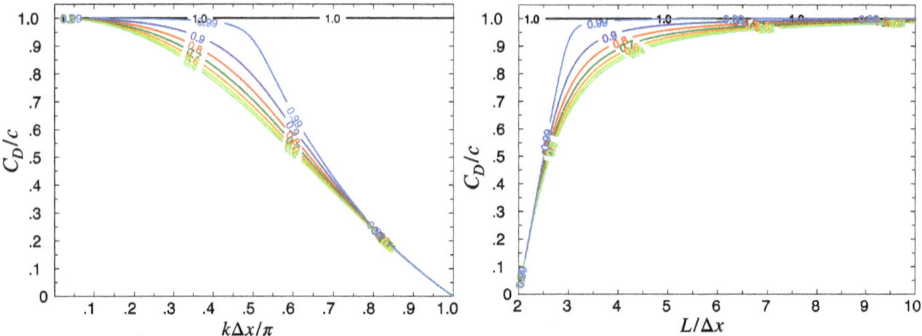

Figure 6.4. Computational dispersion of the leap-frog scheme. The curves show the ratio C_D/c as a function of the normalised wave number $k\Delta x/\pi$ to the left and as a function of the number of grid lengths Δx per wave length L to the right. The different curves correspond to the Courant numbers ($\mu \equiv c\Delta t/\Delta x$) indicated on the them. The ideal solution (black line) is $C_D/c = 1$. Note that when the wave number increases (*i.e.* the wavelength decreases), the computational phase speed deviates from the analytical phase speed. The phase speed is clearly dispersive since the waves propagate at different speeds depending on their wavelengths.

to study the ratio between the analytical and computational group velocities:

$$\frac{C_{Dg}}{c_g} = \frac{\cos\left(k\Delta x\right)}{\sqrt{1 - \left[\mu \sin\left(k\Delta x\right)\right]^2}} \tag{6.5}$$

This ratio is shown in Figure 6.5.

We have encountered two effects in this chapter. The computational phase speed was found to be slower than the analytical phase speed

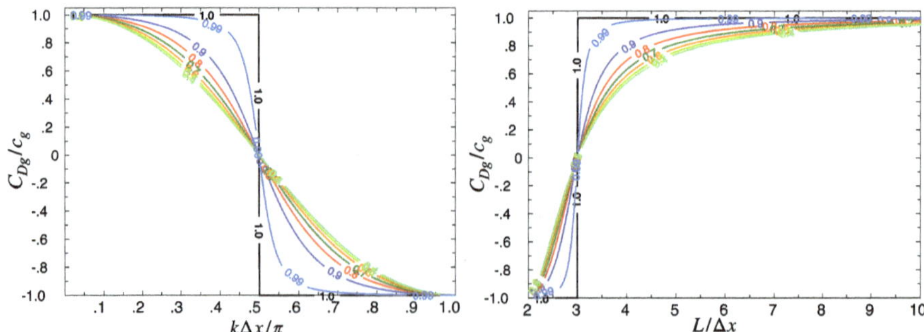

Figure 6.5. As in Figure 6.4 but for the group velocity.

and changes with the wave number; the wave is hence dispersive. The same holds true for the computational group velocity.

Exercise:

Derive the computational phase speed

$$C_D = \frac{1}{k\Delta t} \arcsin\left[\frac{c\Delta t}{\Delta x}\sin\left(k\Delta x\right)\right].$$

7. The Shallow-Water Equations

In this chapter we will consider the equations describing the horizontal propagation of gravity and inertia-gravity waves. These equations are often referred to as the linearised shallow-water equations. We will be dealing with a system of two or three partial differential equations of first order and have two or three dependent variables (one or two velocities and pressure/free-surface height). The system of equations will always be equivalent to a single differential equation, but one of a higher order (which can be obtained from the system by elimination of dependent variables).

7.1 The one-dimensional shallow-water equations

We start with the simplest possible set-up of the shallow-water equations for a flat bottom, *viz.* that associated with gravity waves:

$$\frac{\partial u}{\partial t} = -g\frac{\partial h}{\partial x}, \tag{7.1a}$$

$$\frac{\partial h}{\partial t} = -H\frac{\partial u}{\partial x}, \tag{7.1b}$$

where g is the gravity, h is the thickness of the fluid and H its unperturbed value, *i.e.* when $u = 0$. We seek wave solutions of the form

$$u\left(x, t\right) = u_0 e^{i(kx - \omega t)}, \tag{7.2a}$$

$$h\left(x, t\right) = h_0 e^{i(kx - \omega t)}, \tag{7.2b}$$

which yield the frequency equation

$$\omega^2 = gHk^2,$$

How to cite this book chapter:
Döös, K., Lundberg, P., and Campino, A. A. 2022. *Basic Numerical Methods in Meteorology and Oceanography*, pp. 57–82. Stockholm: Stockholm University Press. DOI: https://doi.org/10.16993/bbs.g. License: CC BY 4.0

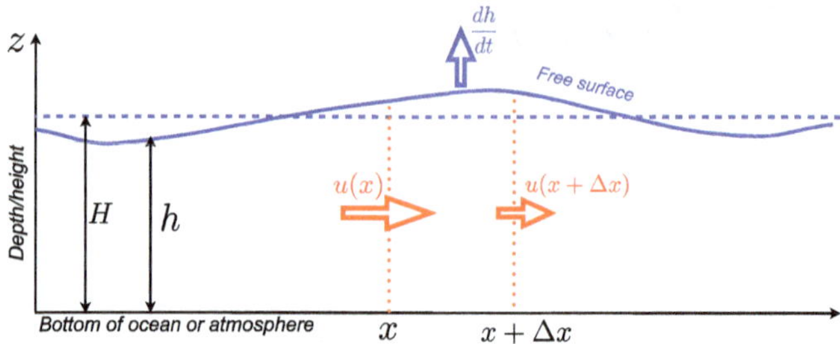

Figure 7.1. Schematic illustration of the 1D shallow-water equations for a flat bottom.

and hence the phase speed is

$$c = \frac{\omega}{k} = \pm\sqrt{gH}.$$

This shows that the gravity waves can propagate along the x-axis in both directions with the speed $\sqrt{gH}$, which is not a function of the wave number, and consequently the waves are non-dispersive.

7.1.1 Spatial discretisation but continuous time derivatives

As illustrated by Figure 7.2, there are two types of possible grids for these shallow-water equations. We can take the two dependent variables at the same points:

$$\frac{\partial u_j}{\partial t} = -g\frac{h_{j+1} - h_{j-1}}{2\Delta x}, \tag{7.3a}$$

$$\frac{\partial h_j}{\partial t} = -H\frac{u_{j+1} - u_{j-1}}{2\Delta x}. \tag{7.3b}$$

This is known as a unstaggered grid. It is also possible to alternate the grid points in space:

$$\frac{\partial u_j}{\partial t} = -g\frac{h_{j+1} - h_j}{\Delta x}, \tag{7.4a}$$

$$\frac{\partial h_j}{\partial t} = -H\frac{u_j - u_{j-1}}{\Delta x}. \tag{7.4b}$$

This is known as a staggered grid and already at this stage we can see that one advantage is that it reduces the number of grid points for a fixed truncation error.

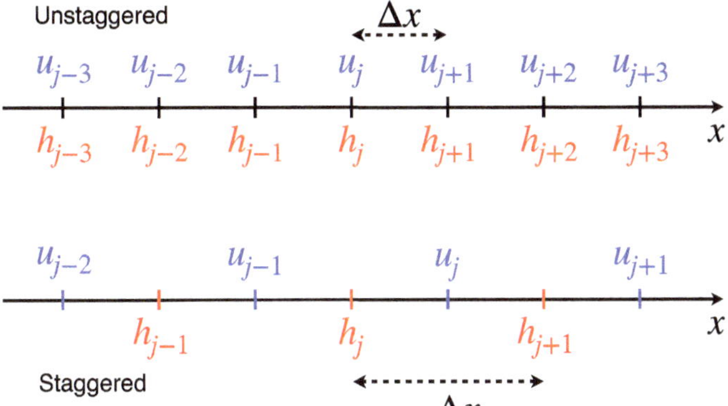

Figure 7.2. Top: unstaggered grid with two dependent variables, both at every grid point. Bottom: staggered grid with two dependent variables at alternate grid points.

The computational phase speeds and group velocities associated with these two schemes can be obtained by inserting wave solutions of the form

$$u_j = u_0 e^{i\left(jk\Delta x - \omega_D t\right)},$$

$$h_j = h_0 e^{i\left(jk\Delta x - \omega_D t\right)}.$$

The computational phase speed derived from the unstaggered-grid equations (7.3) becomes

$$c_D = \frac{\omega_D}{k} = \pm\sqrt{gH}\,\frac{\sin\left(k\Delta x\right)}{k\Delta x} \tag{7.6}$$

and the group velocity is

$$C_{Dg} = \frac{d\left(\omega_D\right)}{dk} = \frac{d\left(kC_D\right)}{dk} = \pm\sqrt{gH}\,\cos\left(k\Delta x\right). \tag{7.7}$$

These are hence the same as for the spatially discretised advection equation in the previous chapter. The corresponding results from the staggered-grid equations (7.4) become

$$c_D = \frac{\omega_D}{k} = \pm\sqrt{gH}\,\frac{\sin\left(k\Delta x/2\right)}{\left(k\Delta x/2\right)} \tag{7.8}$$

and

$$C_{Dg} = \frac{d\left(\omega_D\right)}{dk} = \frac{d\left(kC_D\right)}{dk} = \pm\sqrt{gH}\,\cos\left(\frac{k\Delta x}{2}\right). \tag{7.9}$$

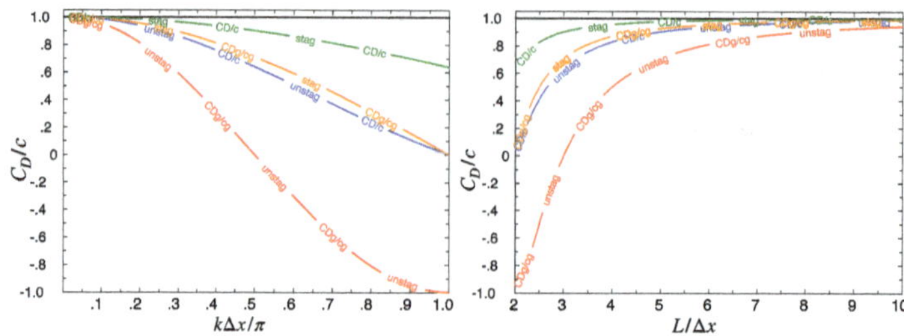

Figure 7.3. The computational phase speed C_D associated with the centred finite-difference scheme in space compared to the analytical phase speed c for the one-dimensional shallow-water equations as a function of the normalised wave number $k\Delta x/\pi$ to the left and as a function of the number of grid lengths Δx per wavelength L to the right. The black line represents the solution of the continuous equations which is the nondispersive analytical case, *i.e.* the phase speed is the same as the group velocity $c = c_g$. The blue curve is the computational phase speed normalised by dividing with c. The red curve shows the computational group velocity C_{Dg} which at wavelengths shorter than 4 grid lengths ($k\Delta x < \pi/2$) propagates in the wrong direction. The green and orange curves are from the staggered grid.

These phase speeds and group velocities are shown and inter-compared in Figure 7.3. Note that when the wave number increases (*i.e.* the wavelength decreases), the computational phase speed deviates from the analytical phase speed. The phase speed is clearly dispersive since the waves propagate at different speeds depending on their wavelengths. Apart from this the staggered grid has the advantage of reducing the number of grid points, and we observe that the waves with $k\Delta x > \pi/2$, which are those shorter than 4 grid lengths and have the largest phase-speed error, are eliminated.

7.1.2 Spatial and temporal discretisation

Equations (7.4) and (7.3) also need to be discretised in time in order to be amenable to a numerical solution. The most straightforward type of time differencing is the three-level leap-frog scheme.

Unstaggered-grid case

Discretisation of the 1D shallow-water equations using centred schemes in both time and space on an unstaggered grid yields

$$\frac{u_j^{n+1} - u_j^{n-1}}{2\Delta t} = -g\frac{h_{j+1}^n - h_{j-1}^n}{2\Delta x}, \tag{7.10a}$$

$$\frac{h_j^{n+1} - h_j^{n-1}}{2\Delta t} = -H\frac{u_{j+1}^n - u_{j-1}^n}{2\Delta x}. \tag{7.10b}$$

We seek solutions of the form

$$u_j^n = u_0 e^{i\left(jk\Delta x - \omega_D n\Delta t\right)},$$

$$h_j^n = h_0 e^{i\left(jk\Delta x - \omega_D n\Delta t\right)},$$

which after insertion in Equations (7.10) yields the computational phase speed

$$C_D = \frac{\omega_D}{k} = \pm\frac{1}{k\Delta t}\arcsin\left[\mu\sin\left(k\Delta x\right)\right], \tag{7.12}$$

where $\mu \equiv \sqrt{gH}\,\Delta t/\Delta x$ is the Courant number. The ratio between the computational and analytical phase speeds should ideally be as close as possible to one, but is

$$\frac{C_D}{c} = \pm\frac{C_D}{\sqrt{gH}} = \pm\frac{1}{k\mu\Delta x}\arcsin\left[\mu\sin\left(k\Delta x\right)\right], \tag{7.13}$$

which is identical to Equation (6.1) found for the advection equation. The computational group velocity is also the same as for the advection equation with the following ratio between the computational and analytical group velocities:

$$\frac{C_{Dg}}{\sqrt{gH}} = \frac{\cos\left(k\Delta x\right)}{\sqrt{1 - \left[\mu\sin\left(k\Delta x\right)\right]^2}}. \tag{7.14}$$

Both computational speeds are functions of the wave number, and thus we recognise that the spatial differencing again results in computational dispersion, *viz.* the same result as obtained for the advection equation with centred schemes.

Staggered-grid case

The same procedure as used above, but on a staggred grid, results in

$$\frac{u_j^{n+1} - u_j^{n-1}}{2\Delta t} = -g\frac{h_{j+1}^n - h_j^n}{\Delta x}, \tag{7.15a}$$

$$\frac{h_j^{n+1} - h_j^{n-1}}{2\Delta t} = -H\frac{u_j^n - u_{j-1}^n}{\Delta x}, \tag{7.15b}$$

which after inserting wave solutions yield the computational phase speed

$$C_D = \frac{\omega_D}{k} = \pm\frac{1}{k\Delta t}\arcsin\left[2\mu\sin\left(k\Delta x/2\right)\right]. \tag{7.16}$$

The ratio between the computational and analytical phase speeds should ideally be as close as possible to one, but is

$$\frac{C_D}{c} = \pm\frac{C_D}{\sqrt{gH}} = \pm\frac{1}{k\mu\Delta x}\arcsin\left[2\mu\sin\left(k\Delta x/2\right)\right],\qquad(7.17)$$

which is shown in the two lower panels of Figure 7.4 together with the results from the unstaggered-grid case in the upper panels.

The ratio between the computational and analytical group velocities for the staggered-grid case becomes

$$\frac{C_{Dg}}{\sqrt{gH}} = \frac{\cos\left(k\Delta x/2\right)}{\sqrt{1-\left[2\mu\sin\left(k\Delta x/2\right)\right]^2}},\qquad(7.18)$$

which clearly differs from that obtained in the unstaggered-grid case.

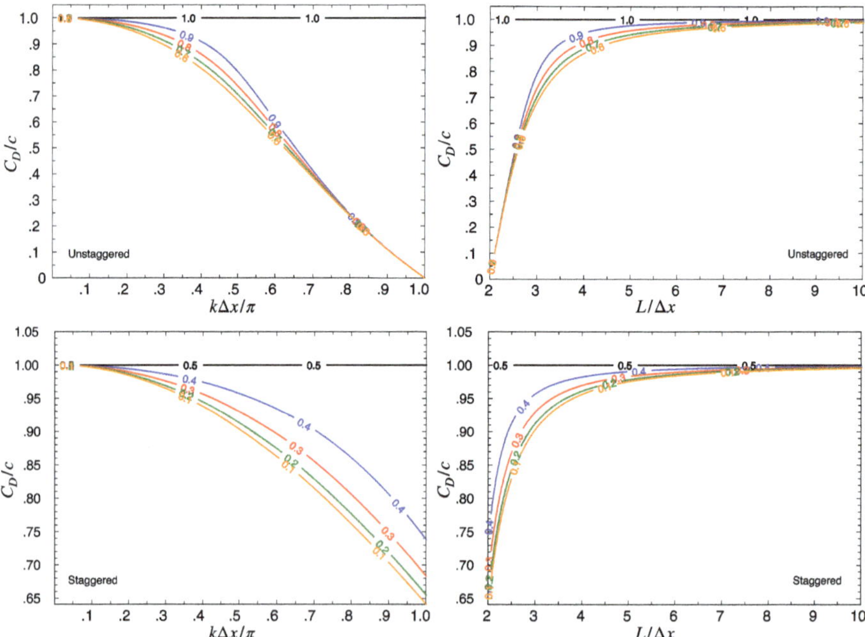

Figure 7.4. The computational phase speed of the 1D shallow-water equations discretised with centred differences in time. The two upper panels represent the unstaggered-grid case and the lower panels the staggered-grid one. The curves show the ratio C_D/c as functions of the normalised wave number $k\Delta x/\pi$ to the left and as functions of the number of grid lengths Δx per wavelength L to the right. The different curves correspond to the Courant numbers μ indicated on them.

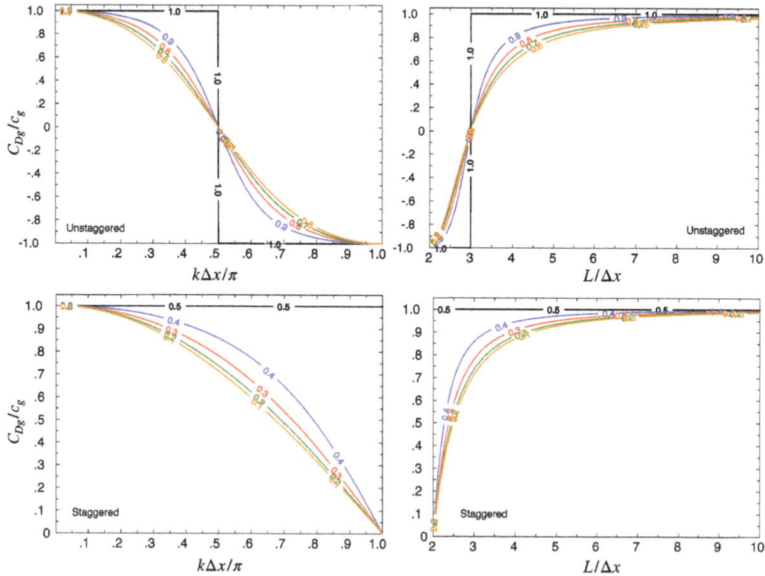

Figure 7.5. As in Figure 7.4 but representing the ratio of the group velocities from Equations (7.14) and (7.18). The two upper panels represent the unstaggered-grid case and the lower the staggered one. The curves show the ratio C_{Dg}/c as a function of the normalised wave number $k\Delta x/\pi$ to the left and as a function of the number of grid lengths Δx per wavelength L to the right. The different curves correspond to the Courant numbers μ indicated on them.

Stability analysis

The stability of the staggered-grid set of equations above can be determined by applying a von Neumann stability analysis, which yields

$$\frac{\lambda - \lambda^{-1}}{2\Delta t} u_0 = -g \frac{e^{ik\Delta x} - 1}{\Delta x} h_0, \tag{7.19a}$$

$$\frac{\lambda - \lambda^{-1}}{2\Delta t} h_0 = -H \frac{1 - e^{-ik\Delta x}}{\Delta x} u_0. \tag{7.19b}$$

We eliminate u_0 and h_0 between these two equations and find that

$$\left(\frac{\lambda - \lambda^{-1}}{2\Delta t}\right)^2 = \frac{gH}{(\Delta x)^2} \left(e^{ik\Delta x} - 1\right)\left(1 - e^{-ik\Delta x}\right), \tag{7.20}$$

which results in two quadratic equations:

$$\lambda^2 \pm 2i\alpha\lambda - 1 = 0, \tag{7.21}$$

where $\alpha \equiv 2\Delta t \sqrt{gH}/\Delta x \sin\left(k\Delta x/2\right)$. The corresponding four roots are:

$$\lambda_{1,2,3,4} = \pm i\alpha \pm \sqrt{1 - \alpha^2}. \tag{7.22}$$

The requirement for stability is that $|\lambda| \leq 1$, which is satisfied if $\alpha < 1$, corresponding to

$$\frac{\Delta t \sqrt{gH}}{\Delta x} \sin\left(\frac{k\Delta x}{2}\right) \leq \frac{1}{2}. \tag{7.23}$$

To satisfy the stability criterion for all wavelengths, we require the Courant number μ to fulfill

$$\mu = \frac{\sqrt{gH}\,\Delta t}{\Delta x} \leq \frac{1}{2}. \tag{7.24}$$

Exercise:

Show that the stability criterion for the unstaggered-grid case is that the Courant number must satisfy $\mu \leq 1$.

7.2 Two-dimensional shallow-water equations

Let us now consider one of the simplest possible subsets of the equations of motion in the atmosphere or the ocean, *viz.* the linearised shallow-water equations (frequently denoted the inertia-gravity wave equations) in two dimensions:

$$\frac{\partial u}{\partial t} - fv = -g\frac{\partial h}{\partial x}, \tag{7.25a}$$

$$\frac{\partial v}{\partial t} + fu = -g\frac{\partial h}{\partial y}, \tag{7.25b}$$

$$\frac{\partial h}{\partial t} = -H\left(\frac{\partial u}{\partial x} + \frac{\partial v}{\partial y}\right). \tag{7.25c}$$

Here $f \equiv 2\Omega \sin\varphi$ is the Coriolis acceleration, where Ω is the angular frequency of the Earth's rotation and φ the latitude. In what follows f is set to be a constant. As before, we seek wave-type solutions:

$$(u, v, h) = (u_0, v_0, h_0)\, e^{i(kx+ly-\omega t)}. \tag{7.26}$$

Insertion into Equations (7.25) yields the following result for the frequency:

$$\omega^2 = f^2 + gH\left(k^2 + l^2\right), \tag{7.27}$$

which describes the dispersion relationship for Poincaré waves (inertia-gravity waves).

7.3 Gravity waves with centred spatial differencing

There are several grids known as "Arakawa grids", which are usually identified by the letters A to E (Mesinger and Arakawa, 1976). The three most commonly used are illustrated in Figure 7.6.

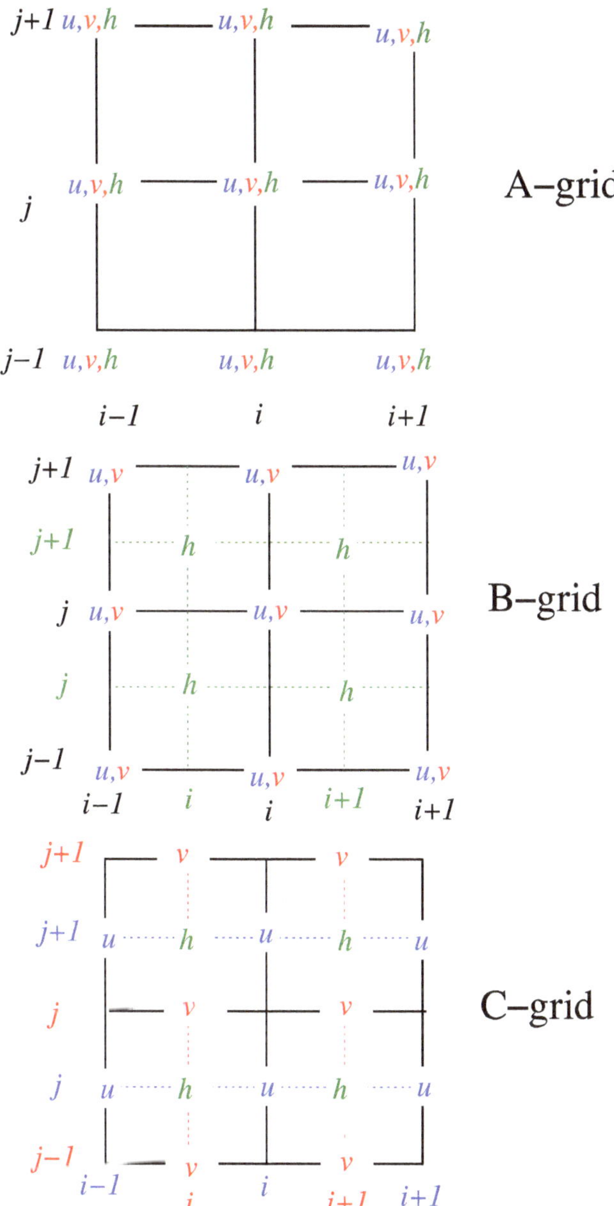

Figure 7.6. The three most common Arakawa grids: A, B and C.

For each of these three grids we use the simplest possible centred approximations for the spatial derivatives, and when necessary for the Coriolis terms. We do not need to study the time differencing since this has previously been examined and remains unchanged.

A-grid:

$$\frac{\partial u_{i,j}}{\partial t} = -g\frac{h_{i+1,j} - h_{i-1,j}}{2\Delta x} + f v_{i,j}, \tag{7.28a}$$

$$\frac{\partial v_{i,j}}{\partial t} = -g\frac{h_{i,j+1} - h_{i,j-1}}{2\Delta y} - f u_{i,j}, \tag{7.28b}$$

$$\frac{\partial h_{i,j}}{\partial t} = -H\left(\frac{u_{i+1,j} - u_{i-1,j}}{2\Delta x} + \frac{v_{i,j+1} - v_{i,j-1}}{2\Delta y}\right). \tag{7.28c}$$

B-grid:

$$\frac{\partial u_{i,j}}{\partial t} = -g\frac{h_{i+1,j} + h_{i+1,j+1} - h_{i,j} - h_{i,j+1}}{2\Delta x} + f v_{i,j}, \tag{7.29a}$$

$$\frac{\partial v_{i,j}}{\partial t} = -g\frac{h_{i,j+1} + h_{i+1,j+1} - h_{i,j} - h_{i+1,j}}{2\Delta y} - f u_{i,j}, \tag{7.29b}$$

$$\frac{\partial h_{i,j}}{\partial t} = -H\Big(\frac{u_{i,j} + u_{i,j-1} - u_{i-1,j} - u_{i-1,j-1}}{2\Delta x}$$
$$+ \frac{v_{i,j} + v_{i-1,j} - v_{i,j-1} - v_{i-1,j-1}}{2\Delta y}\Big). \tag{7.29c}$$

C-grid:

$$\frac{\partial u_{i,j}}{\partial t} = -g\frac{h_{i+1,j} - h_{i,j}}{\Delta x} + \frac{f}{4}\left(v_{i,j} + v_{i+1,j} + v_{i+1,j-1} + v_{i,j-1}\right), \tag{7.30a}$$

$$\frac{\partial v_{i,j}}{\partial t} = -g\frac{h_{i,j+1} - h_{i,j}}{\Delta y} - \frac{f}{4}\left(u_{i,j} + u_{i,j+1} + u_{i-1,j+1} + u_{i-1,j}\right), \tag{7.30b}$$

$$\frac{\partial h_{i,j}}{\partial t} = -H\left(\frac{u_{i,j} - u_{i-1,j}}{\Delta x} + \frac{v_{i,j} - v_{i,j-1}}{\Delta y}\right). \tag{7.30c}$$

For simplicity we shall first study the quasi-one-dimensional case where u, v and h do not depend on y so that Equations (7.25) reduce to

$$\frac{\partial u}{\partial t} - f v = -g\frac{\partial h}{\partial x}, \tag{7.31a}$$

$$\frac{\partial v}{\partial t} + f u = 0, \tag{7.31b}$$

$$\frac{\partial h}{\partial t} + H\frac{\partial u}{\partial x} = 0. \tag{7.31c}$$

Inserting the wave solutions from Equation (7.26) into Equations (7.31) results in the frequency equation

$$\left(\frac{\omega}{f}\right)^2 = 1 + \frac{gH}{f^2}k^2. \tag{7.32}$$

Let us now look at the effect of the finite differencing in space for this case. As the variables are assumed not to depend on y, Equations (7.28) for the A-grid reduce to

$$\frac{\partial u_{i,j}}{\partial t} = -g\frac{h_{i+1,j} - h_{i-1,j}}{2\Delta x} + fv_{i,j}, \tag{7.33a}$$

$$\frac{\partial v_{i,j}}{\partial t} = -fu_{i,j}, \tag{7.33b}$$

$$\frac{\partial h_{i,j}}{\partial t} = -\frac{H}{2\Delta x}\left(u_{i+1,j} - u_{i-1,j}\right), \tag{7.33c}$$

and for the B-grid:

$$\frac{\partial u_{i,j}}{\partial t} = -g\frac{h_{i+1,j} + h_{i+1,j+1} - h_{i,j} - h_{i,j+1}}{2\Delta x} + fv_{i,j}, \tag{7.34a}$$

$$\frac{\partial v_{i,j}}{\partial t} = -fu_{i,j}, \tag{7.34b}$$

$$\frac{\partial h_{i,j}}{\partial t} = -\frac{H}{2\Delta x}\left(u_{i,j} + u_{i,j-1} - u_{i-1,j} - u_{i-1,j-1}\right), \tag{7.34c}$$

and for the C-grid:

$$\frac{\partial u_{i,j}}{\partial t} = -g\frac{h_{i+1,j} - h_{i,j}}{\Delta x} + \frac{f}{4}\left(v_{i,j} + v_{i+1,j} + v_{i+1,j-1} + v_{i,j-1}\right), \tag{7.35a}$$

$$\frac{\partial v_{i,j}}{\partial t} = -\frac{f}{4}\left(u_{i,j} + u_{i,j+1} + u_{i-1,j+1} + u_{i-1,j}\right), \tag{7.35b}$$

$$\frac{\partial h_{i,j}}{\partial t} = -\frac{H}{\Delta x}\left(u_{i,j} - u_{i-1,j}\right). \tag{7.35c}$$

Inserting wave solutions with no j-dependence

$$(u_i, v_i, h_i) = (u_0, v_0, h_0)e^{I\left(ik\Delta x - \omega_D t\right)}$$

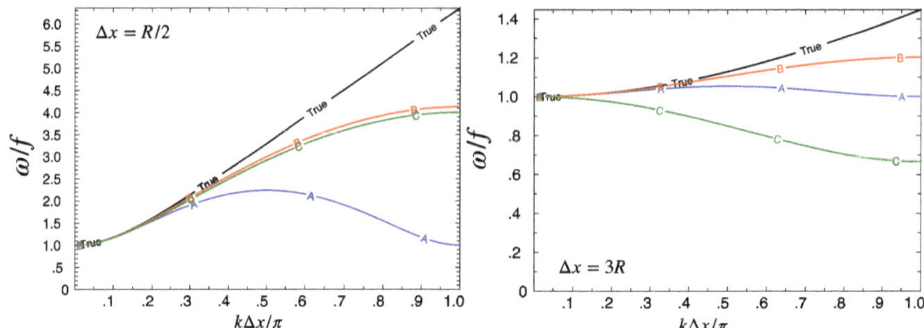

Figure 7.7. The function ω/f from Equation (7.32), where in the left panel $gH/\left(f\Delta x\right)^2 = 4$, *i.e.* the Rossby radius is set to two grid lengths $\left(\sqrt{gH}/f = 2\Delta x\right)$ and in the right panel $gH/\left(f\Delta x\right)^2 = 1/9$, *i.e.* the Rossby radius is set to a third of a grid length $\left(\sqrt{gH}/f = \Delta x/3\right)$. The black curve corresponds to the analytical solution, the blue curve to the results from the A-grid, the red curve to those from the B-grid and the green curve to those from the C-grid. NB: The B- and C-grids yield similar results when the Rossby radius R is well resolved, but the C-grid results degenerate when the grid resolution is coarse.

(where the imaginary unit is denoted I to distinguish it from the spatial index i) into Equations (7.33-7.35) yields the following frequency equations for the three grids:

$$A\ grid : \left(\frac{\omega_D}{f}\right)^2 = 1 + \frac{gH}{f^2}\frac{\sin^2\left(k\Delta x\right)}{\left(\Delta x\right)^2}, \tag{7.36a}$$

$$B\ grid : \left(\frac{\omega_D}{f}\right)^2 = 1 + \frac{gH}{f^2}\frac{\sin^2\left(k\Delta x/2\right)}{\left(\Delta x/2\right)^2}, \tag{7.36b}$$

$$C\ grid : \left(\frac{\omega_D}{f}\right)^2 = \cos^2\left(\frac{k\Delta x}{2}\right) + \frac{gH}{f^2}\frac{\sin^2\left(k\Delta x/2\right)}{\left(\Delta x/2\right)^2}. \tag{7.36c}$$

The non-dimensional frequencies ω_D/f are now seen to depend on the two parameters $k\Delta x$ and gH/f^2, *viz.* the Rossby radius R squared, and are graphed in Figure 7.7, where they can be validated against results from the non-discretised solution of Equation (7.32).

The pros and cons of the three grids can be summarised as follows:

- A-grid: The frequency reaches a maximum at $k\Delta x = \pi/2$, *i.e.* a wave-length of 4 grid intervals. The group velocity is thus zero for this wavelength. If inertia-gravity waves of approximately this wave number are excited near a point inside the computational region, *e.g.* by non-linear effects or by forcing through heating or topography, the wave energy remains near that point. Beyond this maximum value, for $\pi/2 < k\Delta x < \pi$, the frequency decreases as the wave number increases. For these waves the group velocity thus has the wrong sign. Finally, the two-grid-length wave with $k\Delta x = \pi$ behaves like a pure inertial oscillation, and its group velocity is again zero.

- B-grid: The frequency increases monotonically over the range $0 < k\Delta x < \pi$. It, however, assumes a local maximum at the end of the range, and hence the group velocity is zero for the two-grid-length wave with $k\Delta x = \pi$.

- C-grid: If $gH/\left(f\Delta x\right)^2 > 1/4$, the frequency increases monotonically in a similar way as in the B-grid case, *i.e.* when the Rossby radius is larger than half a grid length $\left(\sqrt{gH}/f > \Delta x/2\right)$. If, however, the Rossby radius is exactly half a grid length$\left(\sqrt{gH}/f = \Delta x/2\right)$, the group velocity is zero and for smaller Rossby radii the frequency will decrease in an unrealistic way with increasing wave number over $0 < k\Delta x < \pi$. The advantage of the C-grid lies in that the velocities are normal to the grid-box faces, which makes the differencing of the continuity equation as well as of the scalar transport in the tracer equation a straightforward matter (cf. Section 13.3).

7.4 The shallow-water equations with leap-frog

The discretised linearised inviscid shallow-water equations can now be written with centred finite differences in both time and space on a C-grid as

$$\frac{u_{i,j}^{n+1} - u_{i,j}^{n-1}}{2\Delta t} = -g\frac{h_{i+1,j}^n - h_{i,j}^n}{\Delta x} + \frac{f}{4}\left(v_{i,j}^n + v_{i+1,j}^n + v_{i+1,j-1}^n + v_{i,j-1}^n\right),$$

$$\text{(7.37a)}$$

$$\frac{v_{i,j}^{n+1} - v_{i,j}^{n-1}}{2\Delta t} = -g\frac{h_{i,j+1}^n - h_{i,j}^n}{\Delta y} - \frac{f}{4}\left(u_{i,j}^n + u_{i,j+1}^n + u_{i-1,j+1}^n + u_{i-1,j}^n\right),$$

$$\text{(7.37b)}$$

$$\frac{h_{i,j}^{n+1} - h_{i,j}^{n-1}}{2\Delta t} = -H\left(\frac{u_{i,j}^n - u_{i-1,j}^n}{\Delta x} + \frac{v_{i,j}^n - v_{i,j-1}^n}{\Delta y}\right). \qquad \text{(7.37c)}$$

A von Neumann stability analysis can be applied to the non-rotating case, whereby Equations (7.37) become

$$\frac{\lambda - \lambda^{-1}}{2\Delta t} u_0 = -g \frac{e^{Ik\Delta x} - 1}{\Delta x} h_0, \tag{7.38a}$$

$$\frac{\lambda - \lambda^{-1}}{2\Delta t} v_0 = -g \frac{e^{Il\Delta y} - 1}{\Delta y} h_0, \tag{7.38b}$$

$$\frac{\lambda - \lambda^{-1}}{2\Delta t} h_0 = -H \left(\frac{1 - e^{-Ik\Delta x}}{\Delta x} u_0 + \frac{1 - e^{-Il\Delta y}}{\Delta y} v_0 \right). \tag{7.38c}$$

Note that the imaginary unit here is denoted capital "I" in order to distinguish it from the index "i". Equations (7.38a) and (7.38b) can be rewritten as

$$u_0 = -\frac{2g\Delta t \left(e^{Ik\Delta x} - 1\right)}{\Delta x \left(\lambda - \lambda^{-1}\right)} h_0, \tag{7.39a}$$

$$v_0 = -\frac{2g\Delta t \left(e^{Il\Delta y} - 1\right)}{\Delta y \left(\lambda - \lambda^{-1}\right)} h_0, \tag{7.39b}$$

which we insert into Equation (7.38c), resulting in

$$(\lambda - \lambda^{-1})^2 = \tag{7.40}$$

$$= 4gH \left[\left(\frac{\Delta t}{\Delta x}\right)^2 \left(e^{Ik\Delta x} - 1\right) \left(1 - e^{-Ik\Delta x}\right) \right.$$

$$\left. + \left(\frac{\Delta t}{\Delta y}\right)^2 \left(e^{Il\Delta y} - 1\right) \left(1 - e^{-Il\Delta y}\right) \right]$$

$$= 4gH \left[\left(\frac{\Delta t}{\Delta x}\right)^2 \left(e^{Ik\Delta x/2} - e^{-Ik\Delta x/2}\right)^2 \right.$$

$$\left. + \left(\frac{\Delta t}{\Delta y}\right)^2 \left(e^{Il\Delta y/2} - e^{-Il\Delta y/2}\right)^2 \right]$$

$$= -16gH(\Delta t)^2 \left[\frac{\sin^2(k\Delta x/2)}{(\Delta x)^2} + \frac{\sin^2(l\Delta y/2)}{(\Delta y)^2} \right]$$

$$= -16B^2,$$

where

$$B^2 \equiv gH(\Delta t)^2 \left[\frac{\sin^2(k\Delta x/2)}{(\Delta x)^2} + \frac{\sin^2(l\Delta y/2)}{(\Delta y)^2} \right].$$

Taking the square root of Equation (7.40) results in

$$\lambda - \lambda^{-1} = \pm 4IB, \tag{7.41}$$

which is then transformed into a quadratic equation and has the solution

$$\lambda = \pm 2IB \pm \sqrt{-4B^2 + 1}. \tag{7.42}$$

The requirement for stability is that $|\lambda| \leq 1$, which is satisfied if $B \leq 1/2$. If we assume that $\Delta x = \Delta y$, then to satisfy the stability criterion for all wavelengths we require

$$\mu \equiv \frac{\sqrt{gH}\Delta t}{\Delta x} \leq \frac{1}{\sqrt{8}} \approx 0.35. \tag{7.43}$$

The dispersion relationship can be found by considering the one-dimensional case, *viz.* assuming that the waves propagate along the x-axis ($l \equiv 0$), and substituting

$$\lambda = e^{-I\omega_D \Delta t}$$

in Equation (7.40) so that

$$\omega_D = \frac{1}{\Delta t} \arcsin \left[\frac{\sqrt{gH}\Delta t}{\Delta x/2} \sin\left(k\Delta x/2\right) \right]. \tag{7.44}$$

The computational phase speed is once again

$$C_D = \frac{\omega_D}{k} = \pm \frac{1}{k\Delta t} \arcsin \left[2\mu \sin\left(k\Delta x/2\right) \right], \tag{7.45}$$

which is identical to Equation (7.16), which was the computational phase speed derived directly from the 1D non-rotating shallow-water equations on the staggered grid.

7.5 Boundary conditions

There are several types of boundary conditions: Closed boundary conditions, applied at the border points delimiting land/seafloor from the ocean and solid ground from the atmosphere. Open boundary conditions, taken where the model grid covering the domain under consideration ends but the real ocean/atmosphere continues. A model can also have periodic boundary conditions as previously described for the advection equation in Section 4.2.

7.5.1 Closed boundary conditions

The staggered B-grid is well adapted to no-slip boundary conditions, since the velocity points are located at the corners of the computational grid cell. Unlike the C-grid, there are no ambiguities in the way the dynamical boundary condition is imposed at the "corners" of adjacent land masses as shown in Figure 7.8. The drawback is, however, that for a narrow strait in an ocean model, the B-grid requires at least two grid lengths to have a non-zero velocity point as illustrated in Figure 7.8. The B-grid yields, however, a more satisfactory dispersion relationship than the C-grid since it is better at resolving the Rossby radius at coarse resolutions (Batteen and Han, 1981), a feature that makes this staggering technique suitable for coarsely-resolved models.

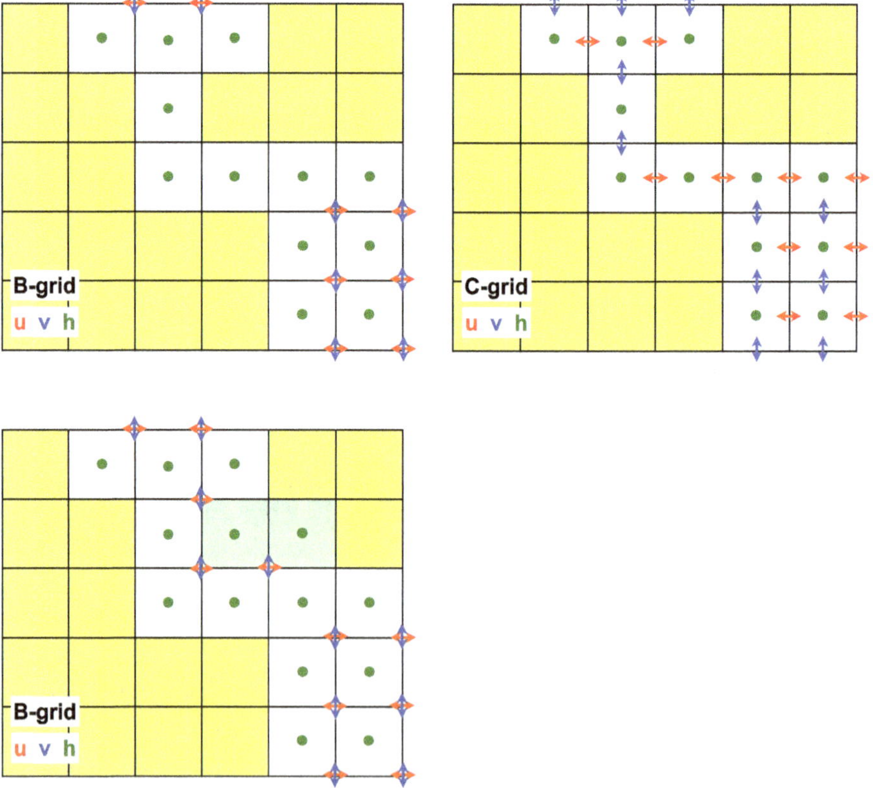

Figure 7.8. Illustration of how a narrow strait is resolved with a B-grid (upper left panel) and with a C-grid (upper right panel). Land cells in yellow, ocean cells in white with corresponding u, v and h points in colour. The only way to permit velocity points in this B-grid strait would be to "dig out" the two cells in green so the strait is at least two grid lengths wide as shown in the lower panel.

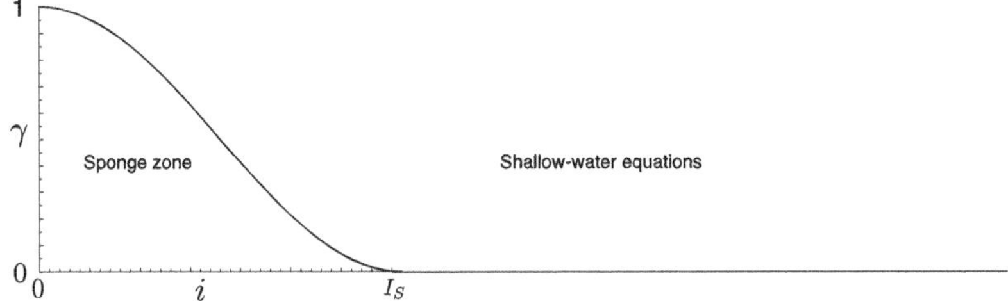

Figure 7.9. Schematic illustration of a sponge zone between the open boundary located at $i = 0$ and $i = I_S$.

7.5.2 Open boundary conditions

An open boundary condition has two main purposes: It should permit waves to propagate out from the model domain without being reflected back. It should also be possible to force the inner solution with external fields, which *e.g.* can be obtained from observations or models covering a larger domain. Open boundary conditions also need to conserve mass so that the average of the sea-surface elevation h remains constant. The energy budget should also be treated accurately, allowing the correct energy flux through the open boundaries to balance the energy flux through the sea surface due to the wind stress.

There are many different types of sophisticated radiative open-boundary conditions based on the wave equation. We will here, however, only present the simplest, which is the "sponge" boundary condition. Here all field variables are first updated using the standard interior leap-frog schemes. The field values in the sponge zone are then relaxed to the externally given values h^E according to

$$h^{n+1} = (1 - \gamma)\, h_*^{n+1} + \gamma h^E, \tag{7.46}$$

where h_*^{n+1} is obtained from the model equations, in the present case the shallow-water equations. The non-uniformity of h^E in the sponge zone is taken into account by letting the solution decay as we leave the boundary. This decay can *e.g.* be of an *e*-folding character or have a cosine-shaped relaxation factor such as

$$\gamma = 0.5 \left[1 + \cos\left(\pi \frac{i}{I_S} \right) \right] \tag{7.47}$$

for the interval $0 \leq i \leq I_S$, where I_S is the number of grid points in the sponge zone, typically 10 to 30.

The externally given values h^E can originate from observations or from another model, which often has a coarser grid. This is the case for regional climate models as well as for most local numerical weather-prediction models, which are forced at their open boundaries by a global circulation model covering the the entire Earth. Figure 7.10 shows schematically such a nested grid, where the light blue border can be treated as an open boundary for the finer interior grid, driven by the values obtained from the coarser-grid model The nesting is hence a way to "zoom" in on a particular region by increasing the spatial resolution here. The nesting can be one-way, where only the interior values are influenced by the exterior values from the coarser surrounding grid or two-way, where also the coarser grid values are affected by the data from the fine grid.

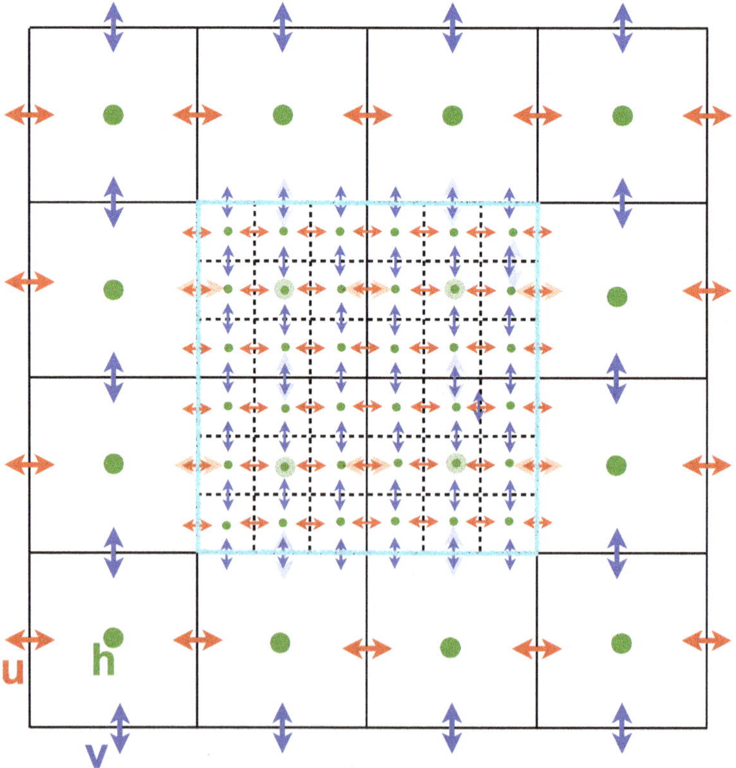

Figure 7.10. A nested C-grid with a 3:1 ratio between the grid sizes. The solid lines denote the coarse grid and the dashed lines represent the fine grid. The light blue lines denote the boundary between the fine and coarse grids, which can be treated as an open boundary for the fine grid.

Radiative boundary conditions, based on the classical Sommerfeld condition (Sommerfeld, 1949), were adapted to oceanic and atmospheric modelling by Orlanski (1976) and followed up by a number of other investigators *e.g.* Higdon (1987) and Flather (1994). These open-boundary conditions have the advantage of letting the waves leave the domain without reflection and at the same time independently imposing the open boundary values h^E. Tests of different radiative boundary conditions can be found in *e.g.* Nycander and Döös (2003).

7.6 Conservation of mass, energy and enstrophy

There are several reasons why numerical schemes for models are often formulated so as to respect conservation properties of the governing equations. An important practical consideration is that satisfying conservation properties helps to ensure the computational stability of a model. Apart from this, the direct physical realism of a conservation property may be a desirable feature. For example, ensuring conservation of mass prevents the surface pressure from drifting to unrealistic values in long-term integrations of atmospheric models. Advection schemes which satisfy an appropriate dynamical conservation property may help to ensure the realism of the simulated energy spectrum. There are, however, considerations other than conservation that might influence the choice of numerical scheme. Shape-preservation (avoidance of the generation of spurious maxima or minima) may be considered as an important feature of an advection scheme, and the economy of a method (especially the ability to accommodate long time steps) may be a critical factor. Indeed, semi-Lagrangian advection schemes (cf. Chapter 11), originally without formal conservation properties, are increasingly being developed for numerical weather prediction.

7.6.1 The shallow-water equations with non-linear advection terms

The shallow-water equations with non-linear advection terms will, following Sadourny (1975), now be presented The momentum equations in vector form can in this case be written as

$$\frac{\partial \vec{V}}{\partial t} + \vec{V} \cdot \nabla \vec{V} + f\vec{k} \times \vec{V} = -g\nabla h, \tag{7.48}$$

which can also be expressed as

$$\frac{\partial \vec{V}}{\partial t} + \xi \vec{k} \times \left(h\vec{V} \right) = -\nabla \left(gh + \frac{1}{2}\vec{V} \cdot \vec{V} \right), \tag{7.49}$$

where $\vec{V}$ is the horizontal velocity vector, f the Coriolis parameter, $\vec{k}$ the unit vector normal to the domain S, $\xi \equiv \left(f + \partial v/\partial x - \partial u/\partial y\right)/h$ the potential vorticity, and h the total water- or air-column height. The continuity equation with non-linear terms is

$$\frac{\partial h}{\partial t} + \nabla \cdot \left(h\vec{V}\right) = 0. \tag{7.50}$$

These equations can also be written in scalar form:

$$\frac{\partial u}{\partial t} - \xi hv = -\frac{\partial B}{\partial x}, \tag{7.51a}$$

$$\frac{\partial v}{\partial t} + \xi hu = -\frac{\partial B}{\partial y}, \tag{7.51b}$$

$$\frac{\partial h}{\partial t} + \frac{\partial\left(hu\right)}{\partial x} + \frac{\partial\left(hv\right)}{\partial y} = 0, \tag{7.51c}$$

where the Bernoulli function is $B \equiv gh + \frac{1}{2}\left(u^2 + v^2\right) = gh + \frac{1}{2}\vec{V}\cdot\vec{V}$. It can be verified that these equations are such that the following properties are conserved:

Total mass: $M = \int_S h\,dS$,

Total Energy: $E = \int_S \frac{1}{2}\left(gh + \vec{V}\cdot\vec{V}\right)h\,dS$,

Absolute potential enstrophy: $Z = \int_S \frac{1}{2}\xi^2 h\,dS$.

Here $\int_S dS$ represents the surface integral over the domain. Enstrophy is the integral of the vorticity squared and can be interpreted as a quantity directly related to the dissipated kinetic energy. By conserving the enstrophy, the model will tend to yield better simulations of eddies.

7.6.2 Discretisation

The discretisation on a C-grid is illustrated in Figure 7.11. The spatial differencing operators δ_x and δ_y acting on u, v and h are

$$\delta_x u = \frac{u_{i,j} - u_{i-1,j}}{\Delta x}; \quad \delta_y v = \frac{v_{i,j} - v_{i,j-1}}{\Delta y}, \tag{7.52}$$

$$\delta_x h = \frac{h_{i+1,j} - h_{i,j}}{\Delta x}; \quad \delta_y h = \frac{h_{i,j+1} - h_{i,j}}{\Delta y}. \tag{7.53}$$

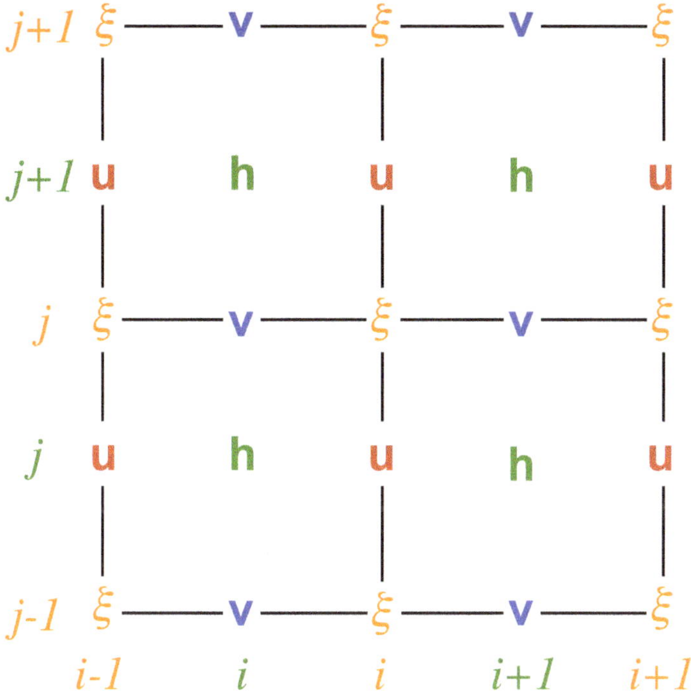

Figure 7.11. C-grid with points for the zonal velocity u, meridional velocity v, water- or air-column height h and vorticity ξ.

Note that all these differences are centred at different points, which is best seen from Figure. 7.11. The spatial averages are similarly

$$\overline{u}^x = \frac{1}{2}\left(u_{i,j} + u_{i-1,j}\right); \ \overline{v}^y = \frac{1}{2}\left(v_{i,j} + v_{i,j-1}\right), \qquad (7.54)$$

$$\overline{h}^x = \frac{1}{2}\left(h_{i,j} + h_{i+1,j}\right); \ \overline{h}^y = \frac{1}{2}\left(h_{i,j} + h_{i,j+1}\right). \qquad (7.55)$$

It is often simpler to use these differencing and averaging operators than to employ index notation. We will use both representations to give the reader an opportunity of suffering the agony of choice.

The mass fluxes U and V are defined at the same points as the velocities u and v:

$$U_{i,j} \equiv \overline{h}^x u = u_{i,j}\frac{1}{2}\left(h_{i,j} + h_{i+1,j}\right),$$

$$V_{i,j} \equiv \overline{h}^y v = v_{i,j}\frac{1}{2}\left(h_{i,j} + h_{i,j+1}\right).$$

The gradient operator will act on the Bernoulli function B defined at the locations where h is defined:

$$B_{i,j} \equiv gh + \frac{1}{2}\left(\overline{u^2}^x + \overline{v^2}^y\right) = gh_{i,j} + \frac{1}{2}\left[\frac{1}{2}\left(u_{i,j}^2 + u_{i-1,j}^2\right)\right.$$
$$\left. + \frac{1}{2}\left(v_{i,j}^2 + v_{i,j-1}^2\right)\right].$$

The potential vorticity is redefined at the corners of the C-grid:

$$\xi_{i,j} = \frac{f + \delta_x v - \delta_y u}{\overline{h}^{xy}} = \frac{f + (v_{i+1,j} - v_{i,j})/\Delta x - (u_{i,j+1} - u_{i,j})/\Delta y}{(h_{i,j} + h_{i+1,j} + h_{i,j+1} + h_{i+1,j+1})/4}.$$

Simple expressions are chosen for a model domain with $N_X \times N_Y$ grid-box faces.

Total mass: $M = \sum\limits_{i=1}^{N_X} \sum\limits_{j=1}^{N_Y} h_{i,j} \Delta x \Delta y$.

Total energy: $E = \frac{1}{2} \sum \left(gh^2 + h\overline{u^2}^x + h\overline{v^2}^y\right) \Delta x \Delta y$

$$= \frac{1}{2} \sum_{i=1}^{N_X} \sum_{j=1}^{N_Y} \left\{gh_{i,j}^2 + \frac{h_{i,j}}{2}\left[\left(u_{i,j}^2 + u_{i-1,j}^2\right) + \left(v_{i,j}^2 + v_{i,j-1}^2\right)\right]\right\} \Delta x \Delta y .$$

Absolute potential enstrophy: $Z = \frac{1}{2} \sum \xi^2 \overline{\overline{h}}^{xy} \Delta x \Delta y$

$$= \frac{1}{2} \sum_{i=1}^{N_X} \sum_{j=1}^{N_Y} \xi_{i,j}^2 \, \frac{1}{4}(h_{i,j} + h_{i+1,j} + h_{i,j+1} + h_{i+1,j+1}) \Delta x \Delta y .$$

Below the symbol $\sum$ refers to a summation of the same species over all grid points. Note that due to symmetry

$$\sum a\overline{b}^x = \sum b\overline{a}^x,$$

and that due to skew-symmetry

$$\sum a\delta_x b = -\sum b\delta_x a.$$

The time derivative of the total energy is

$$\frac{dE}{dt} = \sum \left(U\frac{\partial u}{\partial t} + V\frac{\partial v}{\partial t} + B\frac{\partial h}{\partial t}\right). \tag{7.56}$$

A simple energy-conserving model can be defined as

$$\frac{\partial u}{\partial t} - \overline{\xi \overline{V}^{x}}^{y} + \delta_x B = 0,$$

$$\frac{\partial v}{\partial t} + \overline{\xi \overline{U}^{y}}^{x} + \delta_y B = 0,$$

$$\frac{\partial h}{\partial t} + \delta_x U + \delta_y V = 0.$$

This can also be formulated in a more detailed way using index notation:

$$\frac{\partial u_{i,j}}{\partial t} = \frac{1}{2}\left[\xi_{i,j}\frac{1}{2}\left(V_{i,j} + V_{i+1,j}\right) + \xi_{i,j-1}\frac{1}{2}\left(V_{i,j-1} + V_{i+1,j-1}\right)\right]$$
$$- \frac{B_{i+1,j} - B_{i,j}}{\Delta x}, \tag{7.58a}$$

$$\frac{\partial v_{i,j}}{\partial t} = \frac{1}{2}\left[\xi_{i,j}\frac{1}{2}\left(U_{i,j} + U_{i,j+1}\right) + \xi_{i-1,j}\frac{1}{2}\left(U_{i-1,j} + U_{i-1,j+1}\right)\right]$$
$$- \frac{B_{i,j+1} - B_{i,j}}{\Delta y}, \tag{7.58b}$$

$$\frac{\partial h_{i,j}}{\partial t} = -\frac{U_{i,j} - U_{i-1,j}}{\Delta x} - \frac{V_{i,j} - V_{i,j-1}}{\Delta y}. \tag{7.58c}$$

Energy conservation can be obtained from Equation (7.56):

$$\frac{dE}{dt} = \sum \left(U\overline{\xi \overline{V}^{x}}^{y} - V\overline{\xi \overline{U}^{y}}^{x}\right) + \sum \left(-U\delta_x B - B\delta_x U\right)$$
$$+ \sum \left(-V\delta_y B - B\delta_y V\right) = 0,$$

where each of the three summations cancel out due to the symmetry or skew-symmetry of the operators.

An absolute-potential-enstrophy model can be defined as

$$\frac{\partial u}{\partial t} - \overline{\xi}^{y}\overline{\overline{V}^{x}}^{y} + \delta_x B = 0, \tag{7.59a}$$

$$\frac{\partial v}{\partial t} + \overline{\xi}^{x}\overline{\overline{U}^{y}}^{x} + \delta_y B = 0, \tag{7.59b}$$

$$\frac{\partial h}{\partial t} + \delta_x U + \delta_y V = 0. \tag{7.59c}$$

In the corresponding vorticity equation, the discretised gradients vanish, $viz.$ $\delta_x\delta_y = \delta_y\delta_x$, so that

$$\frac{\partial}{\partial t}\left(\xi\overline{\overline{h}^{x}}^{y}\right) + \delta_x\left(\overline{\xi}^{x}\overline{\overline{U}^{y}}^{x}\right) + \delta_y\left(\overline{\xi}^{y}\overline{\overline{V}^{x}}^{y}\right) = 0,$$

which when combined with the averaged continuity equation

$$\frac{\partial}{\partial t}\left(\overline{h}^{xy}\right) + \delta_x\left(\overline{U}^{yx}\right) + \delta_y\left(\overline{V}^{xy}\right) = 0$$

yields the equation for conservation of the potential enstrophy:

$$\frac{\partial}{\partial t}\left(\xi^2\overline{h}^{xy}\right) + \delta_x\left(\overline{\xi^2}^x\overline{U}^{yx}\right) + \delta_y\left(\overline{\xi^2}^y\overline{V}^{xy}\right) = 0.$$

7.7 A shallow-water model

We will here summarise the results above by formulating a model based on the discretised shallow-water equations on a C-grid. This will be done in the way it is programmed in computer code, and will hence be close to how *e.g.* a Fortran code is structured. The following steps are to be taken:

1. Set the initial condition of the fields for $n = 0$ over the entire model grid indices i and j so that

$$u_{i,j}^{n=0}, \quad v_{i,j}^{n=0}, \quad h_{i,j}^{n=0}$$

   are known.

2. Integrate the shallow-water equations a first time step with an Euler-forward scheme and "loop" over all the model grid indices i and j:

$$u_{i,j}^1 = u_{i,j}^0 + \Delta t\left[-g\frac{h_{i+1,j}^0 - h_{i,j}^0}{\Delta x} + \frac{f}{4}\left(v_{i,j}^0 + v_{i+1,j}^0\right.\right.$$
$$\left.\left. + v_{i+1,j-1}^0 + v_{i,j-1}^0\right)\right],$$

$$v_{i,j}^1 = v_{i,j}^0 + \Delta t\left[-g\frac{h_{i,j+1}^0 - h_{i,j}^0}{\Delta y} - \frac{f}{4}\left(u_{i,j}^0 + u_{i,j+1}^0 + u_{i-1,j+1}^0\right.\right.$$
$$\left.\left. + u_{i-1,j}^0\right)\right],$$

$$h_{i,j}^1 = h_{i,j}^0 - \Delta t H\left(\frac{u_{i,j}^0 - u_{i-1,j}^0}{\Delta x} + \frac{v_{i,j}^0 - v_{i,j-1}^0}{\Delta y}\right).$$

3. Time-integrate the model from $n = 1$ to $n = N_t$, where N_t is the total number of time steps to be computed so that the total time integration will be $N_t \Delta t$. "Leap-frog" the time step with loops over i and j:

$$u_{i,j}^{n+1} = u_{i,j}^{n-1} + 2\Delta t \left[-g\frac{h_{i+1,j}^{n} - h_{i,j}^{n}}{\Delta x} + \frac{f}{4} \left(v_{i,j}^{n} + v_{i+1,j}^{n} \right. \right.$$
$$\left. \left. + v_{i+1,j-1}^{n} + v_{i,j-1}^{n} \right) \right],$$

$$v_{i,j}^{n+1} = v_{i,j}^{n-1} + 2\Delta t \left[-g\frac{h_{i,j+1}^{n} - h_{i,j}^{n}}{\Delta y} - \frac{f}{4} \left(u_{i,j}^{n} + u_{i,j+1}^{n} \right. \right.$$
$$\left. \left. + u_{i-1,j+1}^{n} + u_{i-1,j}^{n} \right) \right],$$

$$h_{i,j}^{n+1} = h_{i,j}^{n-1} - 2\Delta t H \left(\frac{u_{i,j}^{n} - u_{i-1,j}^{n}}{\Delta x} + \frac{v_{i,j}^{n} - v_{i,j-1}^{n}}{\Delta y} \right).$$

4. Apply a Robert-Asselin filter in order to suppress the computational mode:

$$u_{i,j}^{n} = u_{i,j}^{n} + \gamma \left(u_{i,j}^{n-1} - 2u_{i,j}^{n} + u_{i,j}^{n+1} \right),$$
$$v_{i,j}^{n} = v_{i,j}^{n} + \gamma \left(v_{i,j}^{n-1} - 2v_{i,j}^{n} + v_{i,j}^{n+1} \right),$$
$$h_{i,j}^{n} = h_{i,j}^{n} + \gamma \left(h_{i,j}^{n-1} - 2h_{i,j}^{n} + h_{i,j}^{n+1} \right).$$

5. Store the resulting fields at regular time intervals and compute some statistics, *e.g.* the total volume V, the kinetic energy E_P, and available potential energy E_K:

$$V = \sum_{i=1}^{N_X} \sum_{j=1}^{N_Y} h_{i,j}^{n} \Delta x \Delta y,$$

$$E_P = \frac{g}{2} \sum_{i=1}^{N_X} \sum_{j=1}^{N_Y} \left(h_{i,j}^{n} \right)^2 \Delta x \Delta y,$$

$$E_K = \frac{H}{2} \sum_{i=1}^{N_X} \sum_{j=1}^{N_Y} \left[\left(u_{i,j}^{n} \right)^2 + \left(v_{i,j}^{n} \right)^2 \right] \Delta x \Delta y.$$

6. To economise disk space we switch the time-step results (since we only store three of these) before returning to the beginning of the time loop so that $n \to n-1$ and $n+1 \to n$ and

$$u_{i,j}^{n-1} = u_{i,j}^{n}, \quad v_{i,j}^{n-1} = v_{i,j}^{n}, \quad h_{i,j}^{n-1} = h_{i,j}^{n},$$
$$u_{i,j}^{n} = u_{i,j}^{n+1}, \quad v_{i,j}^{n} = v_{i,j}^{n+1}, \quad h_{i,j}^{n} = h_{i,j}^{n+1}.$$

7. End the time loop of the model and the entire model code.

This shallow-water model can be extended to include terms representing non-linear advection and friction/viscosity, the latter to be introduced in next chapter.

8. Diffusion and Friction Terms

In this chapter we will investigate the discretisation of friction and diffusion terms and how this affects the stability of the solution. These terms are included in most models from very simple ones based on the shallow-water equations to highly complex ocean-atmosphere general circulation models.

8.1 Rayleigh friction

We start by studying the simplest type of friction parameterisation, Rayleigh friction, where the retarding acceleration is directly proportional to the velocity. A straightforward example is given by

$$\frac{\partial u}{\partial t} = -\gamma u \,;\, \gamma > 0 \tag{8.1}$$

with the solution

$$u\left(t\right) = u_0 e^{-\gamma t}. \tag{8.2}$$

Using a centred time difference (*i.e.* a leap-frog scheme), which is the most common technique employed for equations with advection terms, the discretisation of Equation (8.1) is

$$\frac{u_j^{n+1} - u_j^{n-1}}{2\Delta t} = -\gamma u_j. \tag{8.3}$$

When a stability analysis is undertaken (in analogy with the one the advection equation was subjected to in the previous chapter) with $u_j^{n+m} = u_j^n \lambda^m$ one finds that

---

How to cite this book chapter:
Döös, K., Lundberg, P., and Campino, A. A. 2022. *Basic Numerical Methods in Meteorology and Oceanography*, pp. 83–93. Stockholm: Stockholm University Press.
DOI: https://doi.org/10.16993/bbs.h. License: CC BY 4.0

1. If the right-hand side of Equation (8.3) is taken at time step n, *i.e.* $-\gamma u_j^n$, then $\lambda_{1,2} = -\gamma\Delta t \pm \sqrt{1 + (\gamma\Delta t)^2}$, which has at least one root that is always greater than one for any $\gamma\Delta t > 0$. The scheme is hence unconditionally unstable.

2. If the right-hand side of Equation (8.3) is taken at time step $n - 1$, *i.e.* $-\gamma u_j^{n-1}$, then $\lambda^2 = 1 - 2\gamma\Delta t$. The scheme is conditionally stable since $\lambda^2 \leqslant 1$ if $\gamma\Delta t \leqslant 1$. But since $\lambda^2 < 0$ for $1/2 < \gamma\Delta t < 1$, the roots of λ will be purely imaginary and the solution u will oscillate and change sign for every second time step. If *e.g.* $\gamma\Delta t = 1$, then $\lambda_{1,2} = \pm i$ and $u^n = i^n = 1, 0, -1, 0, 1, ...$
   For the more restrictive condition $\gamma\Delta t < 1/2$, λ will be real and u will have a more realistic evolution in time with no numerical oscillations.

3. If the right-hand side of Equation (8.3) is taken as an average over the time levels $n - 1$ and $n + 1$, the following finite-difference equation is obtained:

$$\frac{u_j^{n+1} - u_j^{n-1}}{2\Delta t} = -\frac{\gamma}{2}\left(u_j^{n+1} + u_j^{n-1}\right).$$

This, as discussed in Chapter 3, is known as the Crank-Nicolson scheme and is said to be implicit because it includes a term at time level $n + 1$ on its right-hand side. It yields the best approximation of Equation (8.2), and the stability analysis results in $\lambda^2 = (1 - \gamma\Delta t)/(1 + \gamma\Delta t) < 1$. The scheme is hence unconditionally stable. For the same reasons as above one requires $\gamma\Delta t < 1$ in order for a realistic evolution in time. Implicit schemes are often complicated to solve since they include values on both sides of the equation that need to be determined simultaneously. This is, however, not so in this particular case, since the right-hand side is evaluated at the same spatial grid point j as the left-hand side and the equation can be rearranged so that

$$u_j^{n+1} = \frac{1 - \gamma\Delta t}{1 + \gamma\Delta t}\, u_j^{n-1}.$$

When, as in this case, employing the Crank-Nicolson scheme for only a time integration, one should use a two-time-step

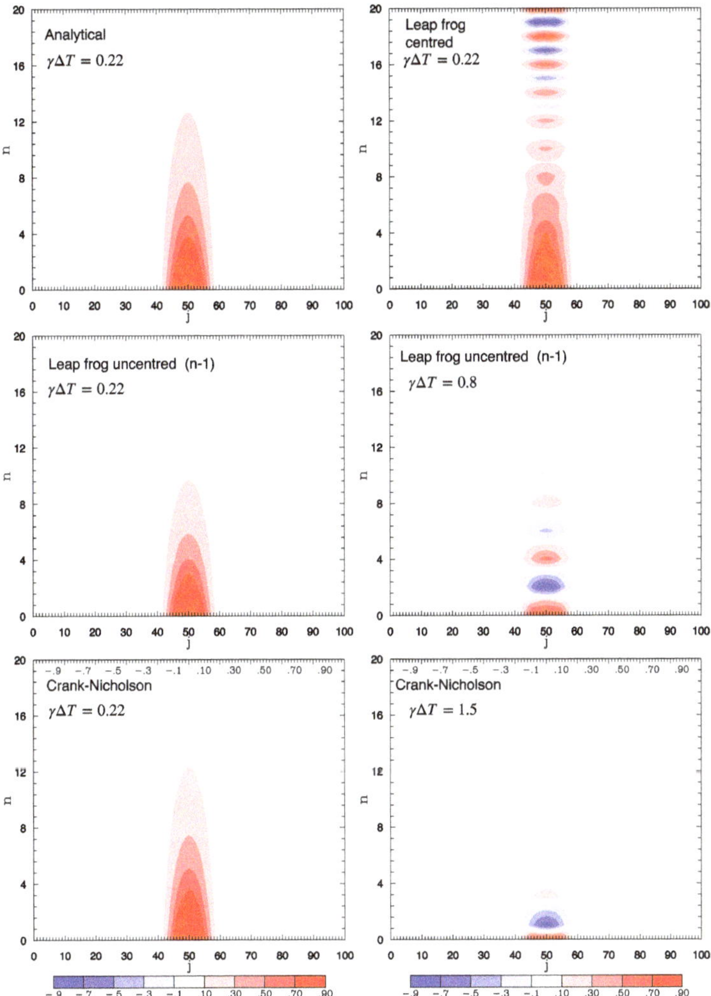

Figure 8.1. The Rayleigh friction equation (8.1) integrated analytically (top left) and numerically with the right-hand side of Equation (8.3) at time step n (top right), which clearly gives an unstable solution. When integrated with the right-hand side at time step $n-1$, the solution is stable and smooth with $\gamma\Delta t = 0.22$ (middle left) but oscillating with $\gamma\Delta t = 0.8$ (middle right). Bottom panels show the results of integration with the Crank-Nicolson scheme giving stable solutions, but oscillating when $\gamma\Delta t = 1.5$.

integration with an Euler-forward scheme so that only two time steps are used and the equation becomes

$$u_j^{n+1} = \frac{1 - \gamma\Delta t/2}{1 + \gamma\Delta t/2}\, u_j^n.$$

The stability analysis above is only strictly valid for these discretisations of the very simple Rayleigh-friction example given by Equation (8.1). However, it turns out that one obtains approximately the same stability criterion when a Rayleigh-friction term is included in the momentum equations or in the tracer equations in a GCM. It is, however, not possible in these cases to undertake a stability analysis of these more comprehensive equations.

8.2 Laplacian friction

A somewhat more realistic friction parameterisation is based on the Laplace operator:

$$\frac{\partial u}{\partial t} = A\frac{\partial^2 u}{\partial x^2}, \tag{8.4}$$

where A is the viscosity coefficient with the unit m^2/s. The letter A originates from the German word *Austausch*, which means "exchange", referring to the exchange of water "parcels". It replaces the molecular viscosity with a much larger eddy viscosity in order to parameterise the sub-grid scales in the momentum equations. Note that often the letter K is used in Equation (8.4), known as the heat equation, when it represents the diffusion of a tracer. This equation is known as the diffusion equation and is a parabolic PDE.

For a single wave number k Equation (8.4) has the solution

$$u(x,t) = u_0 e^{\pm ikx - Ak^2 t}. \tag{8.5}$$

The simplest way to construct a finite-difference approximation of a second-order derivative is to apply finite differencing to a finite difference. This is achieved by first postulating two finite differences centred on the intermediate positions $j + 1/2$ and $j - 1/2$ as illustrated by Figure 8.2:

$$\left(\frac{du}{dx}\right)_{j-1/2} \approx \frac{u_j - u_{j-1}}{\Delta x}, \tag{8.6}$$

$$\left(\frac{du}{dx}\right)_{j+1/2} \approx \frac{u_{j+1} - u_j}{\Delta x}. \tag{8.7}$$

Since the second-order derivative is defined as the derivative of the derivative, we can similarly construct a further finite difference:

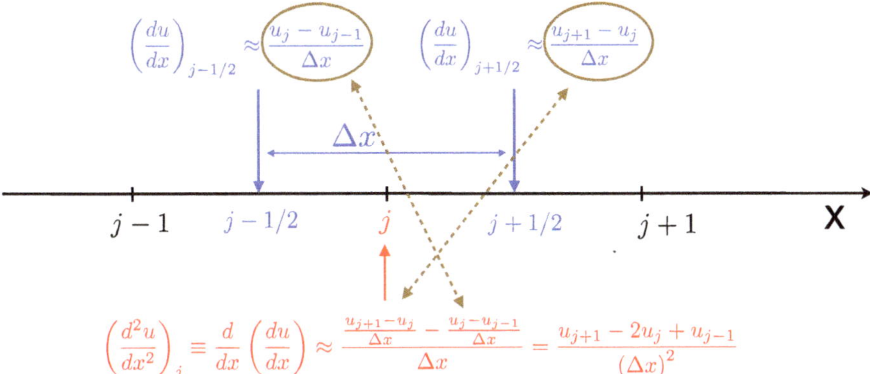

Figure 8.2. The second derivative is the derivative of the derivative. By first estimating the finite differences at $j + 1/2$ and $j - 1/2$ and then the finite difference of those two, one obtains the second finite difference at j.

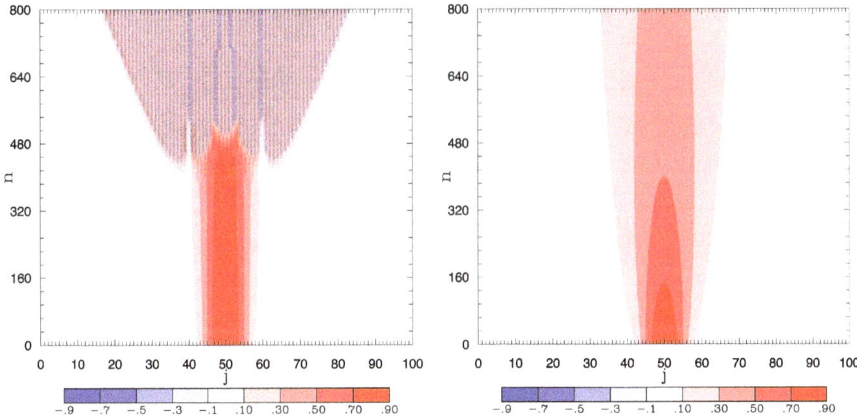

Figure 8.3. The heat equation integrated numerically using Equation (8.12) with the right-hand side at time step n and with $\nu = 0.01$ (left panel) and in the right panel with the right-hand side at time step at $n - 1$ and $\nu = 0.125$.

$$\left(\frac{d^2u}{dx^2}\right)_j \equiv \left[\frac{d}{dx}\left(\frac{du}{dx}\right)\right]_j \simeq \frac{\frac{u_{j+1}-u_j}{\Delta x} - \frac{u_j-u_{j-1}}{\Delta x}}{\Delta x} - \frac{u_{j+1} - 2u_j + u_{j-1}}{(\Delta x)^2}.$$

$$(8.8)$$

The advantage of this formulation is that it is straightforward and intuitively evident. The disadvantage is that it does not provide an estimate of the accuracy of the scheme. To obtain this we use the Taylor-series method previously employed in Section 3.2. A centred finite difference of the Laplace operator corresponding to the second-order derivative can hence be obtained by combining two Taylor series:

$$u_{j+1} = u_j + \Delta x \left(\frac{du}{dx}\right)_j + \frac{(\Delta x)^2}{2}\left(\frac{d^2 u}{dx^2}\right)_j + \frac{(\Delta x)^3}{6}\left(\frac{d^3 u}{dx^3}\right)_j$$
$$+ \frac{(\Delta x)^4}{24}\left(\frac{d^4 u}{dx^4}\right)_j + \cdots, \tag{8.9}$$

$$u_{j-1} = u_j - \Delta x \left(\frac{du}{dx}\right)_j + \frac{(\Delta x)^2}{2}\left(\frac{d^2 u}{dx^2}\right)_j - \frac{(\Delta x)^3}{6}\left(\frac{d^3 u}{dx^3}\right)_j$$
$$+ \frac{(\Delta x)^4}{24}\left(\frac{d^4 u}{dx^4}\right)_j - \cdots. \tag{8.10}$$

Adding these two equations and dividing by $(\Delta x)^2$ we obtain

$$\frac{u_{j+1} - 2u_j + u_{j-1}}{(\Delta x)^2} = \left(\frac{d^2 u}{dx^2}\right)_j + \frac{1}{12}(\Delta x)^2 \left(\frac{d^4 u}{dx^4}\right)_j + \cdots. \tag{8.11}$$

This finite-difference approximation of the second-order derivative is hence accurate to order $(\Delta x)^2$, which is the same as saying it has a second-order truncation error.

The analytical heat equation (8.4) can now be approximated by integrating in time with a leap-frog scheme:

$$\frac{u_j^{n+1} - u_j^{n-1}}{2\Delta t} = A\frac{u_{j+1} - 2u_j + u_{j-1}}{(\Delta x)^2} \tag{8.12}$$

or

$$u_j^{n+1} = u_j^{n-1} + 2\nu \left(u_{j+1} - 2u_j + u_{j-1}\right), \tag{8.13}$$

where $\nu \equiv A\Delta t / (\Delta x)^2$ is the non-dimensional *von Neumann number*, sometimes also called the *diffusion number*. The von Neumann number is, as we will see, now a number that should sometimes not be exceeded in order to have numerical stability when integrating an equation with Laplacian diffusion.

A stability analysis using the von Neumann method is undertaken by inserting $u_j^n = u_0 \lambda^n e^{ikj\Delta x}$ into this equation. Different numerical results are obtained depending on at which time step the right-hand side is chosen. Let us examine the same three cases as we did in the Rayleigh-friction example above:

1. If the right-hand side of Equation (8.12) is taken at time step n, the equation for the amplification factor becomes

$$\lambda^2 + 8\nu \sin^2\left(\frac{k\Delta x}{2}\right)\lambda - 1 = 0,$$

which has the roots $\lambda_{1,2} = -a \pm \sqrt{a^2 + 1}$, where $a \equiv 4\nu \sin^2\left(\frac{k\Delta x}{2}\right)$. For the second root it is recognised that $\lambda_2 < -1$ for any $\nu > 0$, implying that the scheme is unconditionally unstable.

2. If the right-hand side of Equation (8.12) is taken at time step $n - 1$, the amplification-factor equation becomes

$$\lambda^2 = 1 - 8\nu \sin^2\left(\frac{k\Delta x}{2}\right).$$

The scheme is stable when $-1 \leqslant \lambda^2 \leqslant 1$, which is the case when $\nu < 1/4$, and thus the scheme is conditionally stable. However, for the same reasons as for the Rayleigh-friction equation, we recommend the stricter condition $\nu < 1/8$, this in order to have $\lambda^2 > 0$ and hereby avoiding oscillations in time of the solution.

3. If the right hand side of Equation (8.12) is taken as an average of time levels $n - 1$ and $n + 1$ (the Crank-Nicolson scheme) we have

$$\frac{u_j^{n+1} - u_j^{n-1}}{2\Delta t} = \frac{A}{2}\left(\frac{u_{j+1}^{n+1} - 2u_j^{n+1} + u_{j-1}^{n+1}}{(\Delta x)^2} + \frac{u_{j+1}^{n-1} - 2u_j^{n-1} + u_{j-1}^{n-1}}{(\Delta x)^2}\right).$$

$$(8.14)$$

This scheme is implicit as it includes terms at time step $n + 1$ on the right-hand side of the equation, which, however, can not be solved as easily as in the Rayleigh-friction case, this since the $n + 1$ terms on the right-hand side occur at the spatial grid points $j - 1, j, j + 1$. It is, however, possible to use Gaussian elimination to deal with these terms. We can nevertheless undertake a stability analysis and calculate the amplification factor, which is found to be

$$\lambda^2 = \frac{1 - 4\nu \sin^2\left(k\Delta x/2\right)}{1 + 4\nu \sin^2\left(k\Delta x/2\right)}.$$

$$(8.15)$$

Here the right-hand side is always smaller than one and the scheme is hence uncondionally stable. In order to avoid imaginary roots that lead to oscillating solutions one should, however, use $\nu < 1/4$.

In most cases when modelling the atmosphere or the ocean, γ and A are of such magnitudes that the stability criterion derived in the present chapter permits Δt to be much larger than the value conforming to the CFL criterion, which requires a restriction of the Courant number $\mu \equiv c\Delta x/\Delta t$. A common mistake when writing a simple model code is, however, to use the unconditionally unstable scheme with the friction taken at time step n.

Note that the schemes in the two last cases discussed above are in fact two-level schemes, since we do not use any values at time step n, but only at $n-1$ and $n+1$. There is consequently no reason to use a leap-frog scheme here, and we can instead use an Euler-forward scheme in time and replace all time levels $n-1$ by n. The stability analysis remains unaltered, but, since the time step is halved, we should replace Δt by $\Delta t/2$. It is nevertheless easier to demonstrate the differences between the three cases by using leap-frog schemes for all of them.

In Section 4.3 we saw that wave propagation with the discretised advection equation required restrictions on the Courant number μ. Here, we have seen that similar stability conditions arise when Rayleigh and Laplacian friction are used. This leads to restrictions on the non-dimensional number $\gamma\Delta t$ and the von Neumann number $\nu \equiv A\Delta t/(\Delta x)^2$. The discretised momentum equation will need to satisfy all these stability criteria when friction parameterisations are included and the CFL criterion is satisfied. The next section will examine how the stability criteria associated with advection and diffusion are interrelated.

8.3 The advection-diffusion equation

Let us now examine an equation with both advection and diffusion terms:

$$\frac{\partial u}{\partial t} + c\frac{\partial u}{\partial x} = A\frac{\partial^2 u}{\partial x^2}, \tag{8.16}$$

which in this form combines both the parabolic and hyperbolic properties of a partial differential equation. It has the analytical solution

$$u(x,t) = u_0 e^{\pm ik(x-ct)-Ak^2 t}. \tag{8.17}$$

We have previously seen that a discretisation of the advection equation with a scheme centred in time as well as in space is stable, while for the diffusion equation the discretised Laplace operator must be taken at time step $n - 1$ in order to ensure stability. Let us now combine these schemes:

$$\frac{u_j^{n+1} - u_j^{n-1}}{2\Delta t} + c\frac{u_{j+1}^n - u_{j-1}^n}{2\Delta x} = A\frac{u_{j+1}^{n-1} - 2u_j^{n-1} + u_{j-1}^{n-1}}{(\Delta x)^2} \qquad (8.18)$$

or

$$u_j^{n+1} = u_j^{n-1} - \mu\left(u_{j+1}^n - u_{j-1}^n\right) + 2\nu\left(u_{j+1}^{n-1} - 2u_j^{n-1} + u_{j-1}^{n-1}\right). \qquad (8.19)$$

We now undertake a stability analysis by inserting $u_j^n = \lambda^n e^{ikj\Delta x}$ into this equation, which after some calculations yields

$$\lambda^2 + 2ia\lambda + 8b - 1 = 0, \qquad (8.20)$$

with the coefficients $a \equiv \mu\sin\left(k\Delta x\right)$ and $b \equiv \nu\sin^2\left(k\Delta x/2\right)$.

The solution of Equation (8.20) is

$$\lambda = -ia \pm \sqrt{-a^2 + 1 - 8b}.$$

If $1 - 8b - a^2 > 0$ then $|\lambda|^2 = a^2 + 1 - 8b - a^2 = 1 - 8b < 1$, *viz.* the scheme is stable for this root.

If $1 - 8b - a^2 < 0$ then $\lambda = -i\left(a \mp \sqrt{a^2 + 8b - 1}\right) \Rightarrow \lambda^2 = -\left(2a^2 + 8b - 1 \mp 2a\sqrt{a^2 + 8b - 1}\right)$.

It is not immediately evident when the second root of this expression for λ^2 yields a stable solution, and thus we have graphed λ^2 as a function of a and b in Figure 8.4.

The stability analysis above can be tested by reformulating Equation (8.19) as

$$u_j^{n+1} = u_j^{n-1} - \mu\left(u_{j+1}^n - u_{j-1}^n\right) + 2\nu\left(u_{j+1}^{n-1} - 2u_j^{n-1} + u_{j-1}^{n-1}\right), \qquad (8.21)$$

which we then integrate numerically for 100 time steps with the same initial condition as for the Rayleigh and diffusion equations in the previous sections. The choice of both the Courant and von Neumann numbers will determine the stability of the integration. From Figure 8.4 we can see that a stable and non-oscillating solution will require a Courant number μ below 1. If we choose *e.g.* $\nu = 1/16$, we recognise from Figure 8.4 that $|\lambda|^2 < 1$ for μ up to approximately 0.70. To test this we have integrated Equation (8.21) with a Courant number μ just above and

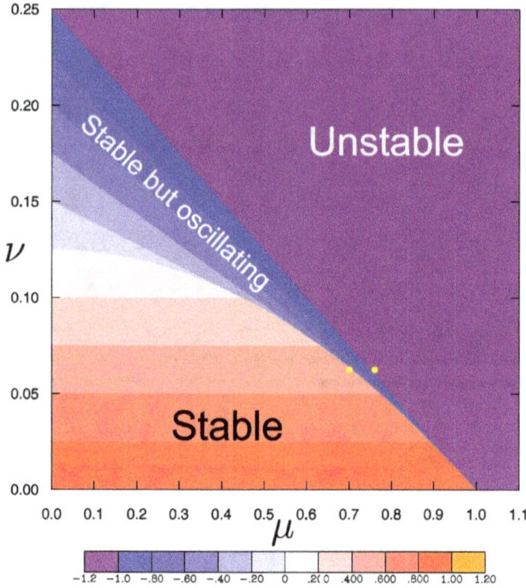

Figure 8.4. The squared amplification factor λ^2 as a function of the Courant number μ and the von Neumann number ν. The purple region, where $\lambda^2 < -1$, corresponds to where the solutions are unstable. The blue region, where $-1 < |\lambda|^2 < 0$, represents stable but oscillating solutions. The red region is for $0 < \lambda^2 < 1$, which corresponds to stable solutions with no oscillations. The two yellow dots indicate $\nu = 1/16$ with $\mu = 0.70$ and $\mu = 0.76$, which are the two test cases illustrated in Figure 8.5.

below this critical value. Figure 8.5 shows the results of these two integrations, with a stable solution obtained for $\mu = 0.70$ and an unstable one for $\mu = 0.76$.

Exercises

1. Undertake a stability analysis for the Rayleigh-friction equation with the right-hand side of Equation (8.3) taken at time step n.
2. Same as in 1) but for the right-hand side taken at time step $n - 1$.
3. Same as in 1) but for the right-hand side taken at time step $n + 1$.
4. Calculate the stability criterion for

$$\frac{\partial u}{\partial t} = A \frac{\partial^2 u}{\partial x^2} \; ; \; A > 0,$$

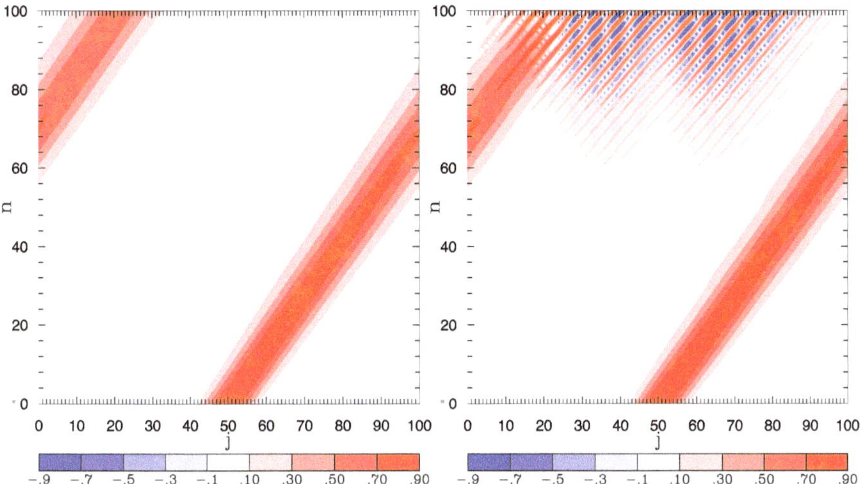

Figure 8.5. The heat-diffusion equation integrated numerically using Equation (8.21) with $\nu = 1/16$ and $\mu = 0.70$ in the left panel and $\mu = 0.76$ in the right panel. Note the visibly growing instabilities in the right panel after 60 time steps.

using the following scheme:

$$\frac{u_j^{n+1} - u_j^n}{2\Delta t} = A\frac{u_{j+1}^n - 2u_j^n + u_{j-1}^n}{(\Delta x)^2}.$$

Estimate an upper limit for Δt when
  i) $A = 10^6 m^2/s$, $\Delta x = 400\,km$ (large-scale horizontal diffusion),
  ii) $A = 1m^2/s$, $\Delta x = 10\,m$ (vertical diffusion in a boundary layer).

5. The diffusion equation can be integrated using the Crank-Nicolson scheme:

$$\frac{T_j^{n+1} - T_j^n}{\Delta t} = \frac{A}{2}\left[\frac{T_{j+1}^n - 2T_j^n + T_{j-1}^n}{(\Delta x)^2} + \frac{T_{i+1}^{n+1} - 2T_j^{n+1} + T_{j-1}^{n+1}}{(\Delta x)^2}\right].$$

Examine the stability of this scheme!

9. The Poisson and Laplace Equations

Consider the elliptic Poisson equation in two dimensions:

$$\nabla^2 u = \left(\frac{\partial^2}{\partial x^2} + \frac{\partial^2}{\partial y^2} \right) u = f(x, y). \tag{9.1}$$

If $f(x, y) = 0$ this is known as the Laplace equation. As shown by Figure 9.1, Equation (9.1) can be discretised on a grid:

$$\frac{u_{i-1,j} - 2u_{i,j} + u_{i+1,j}}{(\Delta x)^2} + \frac{u_{i,j-1} - 2u_{i,j} + u_{i,j+1}}{(\Delta y)^2} = f_{i,j}, \tag{9.2}$$

which can also be written as

$$u_{i,j} = \frac{(\Delta y)^2 \left(u_{i-1,j} + u_{i+1,j} \right) + (\Delta x)^2 \left(u_{i,j-1} + u_{i,j+1} \right) - (\Delta x \Delta y)^2 f_{i,j}}{2 \left[(\Delta x)^2 + (\Delta y)^2 \right]}. \tag{9.3}$$

If we consider a square grid such that $\Delta x = \Delta y$, Equation (9.3) simplifies to

$$u_{i,j} = \frac{1}{4} \left[u_{i-1,j} + u_{i+1,j} + u_{i,j-1} + u_{i,j+1} - (\Delta x)^2 f_{i,j} \right]. \tag{9.4}$$

When the boundary values for the domain are known, it is possible to solve this finite-difference equation by iteration.

In iterative methods we need initial values at iteration level $u_{i,j}^m$ ($m = 0$ initially) and the purpose is to calculate $u_{i,j}^{m+1}$. This procedure is repeated until the difference between the results from two successive iterations decrease below a prescribed value at each grid point.

How to cite this book chapter:
Döös, K., Lundberg, P., and Campino, A. A. 2022. *Basic Numerical Methods in Meteorology and Oceanography*, pp. 95–100. Stockholm: Stockholm University Press. DOI: https://doi.org/10.16993/bbs.i. License: CC BY 4.0

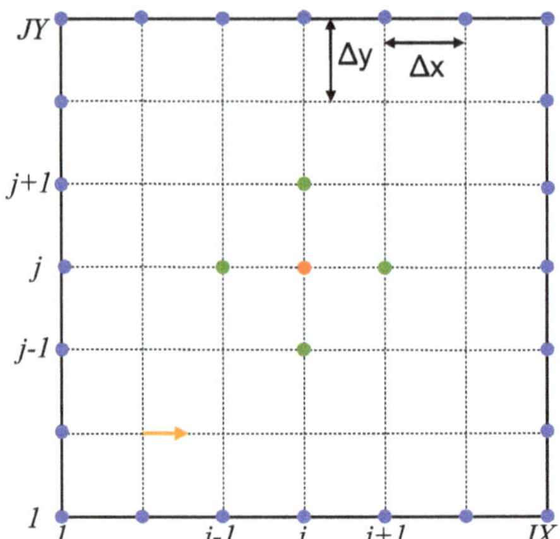

Figure 9.1. Grid for the Poisson and Laplace equations. Boundary values required for the four walls at the blue points. The green points illustrate which points are needed to compute the red point. The solution is obtained by calculating iteratively starting from the orange arrow.

9.1 Jacobi iteration

Values from the previous iteration level are used, which results in

$$u_{i,j}^{m+1} = \frac{1}{4}\left[u_{i-1,j}^m + u_{i+1,j}^m + u_{i,j-1}^m + u_{i,j+1}^m - (\Delta x)^2 f_{i,j}\right]. \qquad (9.5)$$

The method works but is somewhat inefficient and is not used for solving practical problems.

A Fortran-code segment of this iterative computation could look like this:

```fortran
u=0. ! initialise the field to zero
f=0. ! Set the function f to something
omega=1.5 ! relaxation factor
! Set the boundary conditions to something
u(1,:,:)=10. ; u(IX,:,:)=10.
u(:,1,:)=10. ; u(:,JY,:)=10.
do m=1,500 ! number of iterations
do j=2,JY-1
 do i=2,IX-1
!    Jacobi iteration
!    u(i,j,m+1)=0.25*(u(i-1,j,m)+u(i+1,j,m) &
!    &           +u(i,j-1,m)+u(i,j+1,m)-dx**2*f(i,j))
!    Gauss-Seidel iteration
!    u(i,j,m+1)=0.25*(u(i-1,j,m+1)+u(i+1,j,m) &
!    & +u(i,j-1,m+1)+u(i,j+1,m)-dx**2*f(i,j) )
```

```fortran
! Successive Over Relaxation (SOR)
  u(i,j,m+1)=(1.-omega)*u(i,j,m)+ &
  & omega*0.25*(u(i-1,j,m+1)+u(i+1,j,m) + &
  & u(i, j-1,m+1)+u(i,j+1,m)-dx**2*f(i,j) )

  res(m)=res(m)+(u(i,j,m+1)-u(i,j,m))**2
 enddo
 enddo
 print *,m,res(m)
enddo
```

9.2 Gauss-Seidel iteration

A clear improvement in efficiency of iterative methods is obtained if we use the newly computed values in the iteration formula: iteration level $m + 1$ values are available for nodes $(i - 1, j)$ and $(i, j - 1)$ when calculating u for node (i, j). Thus the Gauss-Seidel formula is:

$$u_{i,j}^{m+1} = \frac{1}{4} \left[u_{i-1,j}^{m+1} + u_{i+1,j}^{m} + u_{i,j-1}^{m+1} + u_{i,j+1}^{m} - (\Delta x)^2 f_{i,j} \right]. \qquad (9.6)$$

The inclusion of the two newly computed values makes Gauss-Seidel iteration more efficient than Jacobi iteration. Note that the Fortran code above for the Jacobi method can easily be "upgraded" to the Gauss-Seidel method by simply removing the iteration index m.

9.3 Successive Over Relaxation (SOR)

The Gauss-Seidel iteration method can be further improved by increasing the convergence rate using the method of Successive Over Relaxation (SOR). The change between two successive Gauss-Seidel iterations is denoted the residual c, which is defined as

$$c = u_{i,j}^{m+1} - u_{i,j}^{m}. \qquad (9.7)$$

In the method of SOR, the Gauss-Seidel residual is multiplied by a relaxation factor ω and a new iteration value is obtained from

$$u_{i,j}^{m+1} = u_{i,j}^{m} + \omega c = u_{i,j}^{m} + \omega \left(\hat{u}_{i,j}^{m+1} - u_{i,j}^{m} \right) = (1 - \omega) u_{i,j}^{m} + \omega \hat{u}_{i,j}^{m+1}, \qquad (9.8)$$

where $\hat{u}_{i,j}^{m+1}$ denotes the new iteration value obtained from the Gauss-Seidel method using Equation (9.6). It can easily be seen that if $\omega = 1$, SOR reduces to the Gauss-Seidel iteration method.

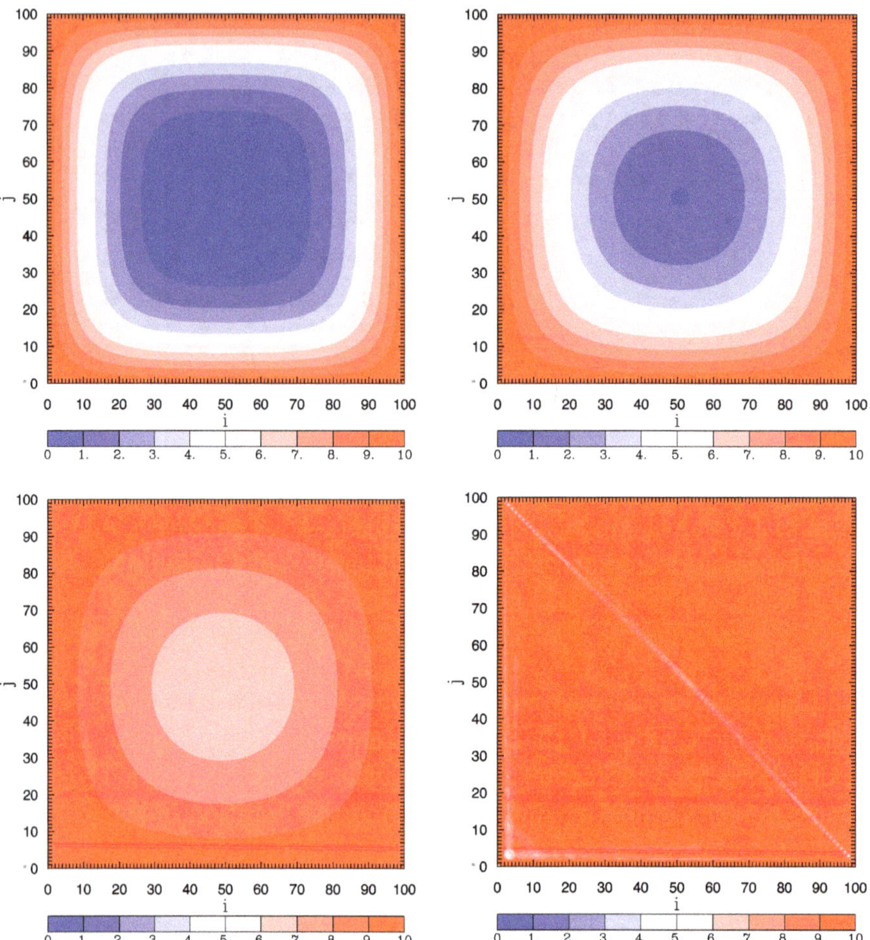

Figure 9.2. Numerical solution of the Laplace equation using the supplied Fortran code. Iterated 500 times using the Jacobi method (upper left), the Gauss-Seidel method (upper right), the SOR method with $\omega = 1.5$ (lower left) and with $\omega = 2$ (lower right). Note that this last case does not converge because of too large a relaxation factor. The end solution should converge towards 10, since that is the value of all the boundary conditions and $f = 0$.

By substituting Equation (9.6) of the Gauss-Seidel iteration method in Equation (9.8), we obtain the equation used in the SOR method:

$$u_{i,j}^{m+1} = (1 - \omega)\, u_{i,j}^{m} + \frac{\omega}{4} \left[u_{i-1,j}^{m+1} + u_{i+1,j}^{m} + u_{i,j-1}^{m+1} + u_{i,j+1}^{m} - (\Delta x)^2 f_{i,j} \right].$$

$$(9.9)$$

Usually the numerical value of the relaxation factor ω can be obtained by trial and error and the optimum value is generally around 1.5. In the case that $0 < \omega < 1$, the method is said to be "under-relaxed". According to the choice of the parameter ω, we either extrapolate for $\omega > 1$ or for $0 < \omega < 1$ interpolate between the old iteration value at level m and the Gauss-Seidel value at level $m + 1$. If we extrapolate too much, *i.e.* ω is taken too large, the iteration starts to oscillate and probably collapses. The iterative methods described above are often referred to as *relaxation* methods as an initially guessed solution is allowed to slowly relax, reducing the errors, towards the true solution.

Finally, it also deserves mention that multi-grid methods can also be employed, where sequences of coarser grids are used so as to provide initial guesses for the finer grids. This is done in order to speed up the convergence of the iterative procedure. See *e.g.* Hackbusch (1985) for a comprehensive overview of multi-grid methods and applications.

9.4 Helmholtz Decomposition

An example of Poisson equations, where one uses the iterative methods above in order to solve the equation, is when computing stream functions and velocity potentials. The Helmholtz theorem states that a velocity field can be decomposed into rotational and divergent parts:

$$u = u_\chi + u_\psi, \tag{9.10a}$$

$$v = v_\chi + v_\psi, \tag{9.10b}$$

where the subscript χ denotes the divergent irrotational part and ψ the non-divergent rotational part. The stream function is defined as

$$\frac{\partial \psi}{\partial x} = v_\psi \, , \frac{\partial \psi}{\partial y} = -u_\psi. \tag{9.11}$$

The velocity potential is defined as

$$\frac{\partial \chi}{\partial x} = u_\chi \, , \frac{\partial \chi}{\partial y} = v_\chi. \tag{9.12}$$

An equation for the stream function ψ can be derived by

$$\frac{\partial 9.10b}{\partial x} - \frac{\partial 9.10a}{\partial y} :$$

$$\frac{\partial v}{\partial x} - \frac{\partial u}{\partial y} = \frac{\partial v_\chi}{\partial x} + \frac{\partial v_\psi}{\partial x} - \frac{\partial u_\chi}{\partial y} - \frac{\partial u_\psi}{\partial y} =$$

$$\frac{\partial^2 \psi}{\partial x^2} + \frac{\partial^2 \chi}{\partial x \partial y} + \frac{\partial^2 \psi}{\partial y^2} - \frac{\partial^2 \chi}{\partial y \partial x} = \nabla^2 \psi = \xi(x, y), \tag{9.13}$$

where ξ is the relative vorticity.

An equation for the velocity potential can similarly be obtained by

$$\frac{\partial 9.10a}{\partial x} + \frac{\partial 9.10b}{\partial y} :$$

$$\nabla^2 \psi = \nabla \cdot \vec{V}. \tag{9.14}$$

These two Poisson equations may be solved iteratively as explained previously in this chapter.

Exercise:

Set up a numerical model of the Laplace equation with a grid of 10 x 10 points. Start with $u = 0$ in the interior and with $u = 1$ as boundary conditions. Test the convergence of the 3 different iteration schemes from this chapter.

10. Implicit and Semi-Implicit Schemes

The time step permitted by the economical explicit schemes, twice that satisfying the CFL criterion, is still considerably shorter than that required for accurate integration of the quasi-geostrophic equations, which do not permit fast oscillating waves. Thus we will here consider implicit schemes, which have the pleasing property of being stable for any choice of time step.

10.1 Implicit versus explicit schemes, a simple example

For implicit schemes, the spatial terms are evaluated, at least partially, at the unknown time level. Let us consider one of the simplest possible examples by examining the one-dimensional diffusion equation, also known as the heat equation (8.4):

$$\frac{\partial u}{\partial t} = A\frac{\partial^2 u}{\partial x^2}. \tag{10.1}$$

A formal solution of this parabolic differential equation requires an initial condition as well as two boundary conditions, the latter, in order not to complicate our problem unnecessarily, here taken to be Dirichlet conditions (cf. Section 2.1)

This equation is discretised with centred spatial finite differences and integrated in time with an Euler-forward scheme. In traditional explicit form we obtain

$$u_i^{n+1} = u_i^n + A\frac{\Delta t}{(\Delta x)^2}\left(u_{i-1}^n - 2u_i^n + u_{i+1}^n\right), \tag{10.2}$$

How to cite this book chapter:
Döös, K., Lundberg, P., and Campino, A. A. 2022. *Basic Numerical Methods in Meteorology and Oceanography*, pp. 101–107. Stockholm: Stockholm University Press. DOI: https://doi.org/10.16993/bbs.j. License: CC BY 4.0

which can be rewritten as

$$u_i^{n+1} = \nu u_{i-1}^n + (1 - 2\nu)u_i^n + \nu u_{i+1}^n, \tag{10.3}$$

where $\nu \equiv A\Delta t/(\Delta x)^2$ is as previously the von Neumann number. This equation is explicit in terms of u_i^{n+1}, which is the value at the unknown time level $n+1$, and is hence possible to solve. A stability analysis can be performed and shows that it is conditionally stable $(-1 \leq \lambda \leq 1)$ for $\nu \leq 1/2$. A stricter condition with only a positive root $(0 \leq \lambda \leq 1)$ for the non oscillating solution is obtained for $\nu \leq 1/4$.

A similar approach, but evaluating the spatial term at the unknown time level $n+1$, yields the fully implicit discretisation

$$u_i^{n+1} = u_i^n + \nu\left[u_{i-1}^{n+1} - 2u_i^{n+1} + u_{i+1}^{n+1}\right], \tag{10.4}$$

which can be rewritten with all terms at time step $n+1$ on the left-hand side as

$$-\nu u_{i-1}^{n+1} + (1 + 2\nu)u_i^{n+1} - \nu u_{i+1}^{n+1} = u_i^n. \tag{10.5}$$

This implicit discretisation is unconditionally stable. To solve the equation one needs to consider all grid points i. In the present case, when we are dealing with the linearised heat equation, the problem can be expressed as a linear system of equations $\mathbf{A}\vec{X} = \vec{B}$, where $\mathbf{A}$ is a matrix, $\vec{X}$ a vector given by the unknown values of u at time $n+1$, and $\vec{B}$ a vector given by the known values of u:

$$\begin{bmatrix} (1+2\nu) & -\nu & & & & \\ -\nu & (1+2\nu) & -\nu & & & \\ \cdots & \cdots & \cdots & & & \\ & \cdots & \cdots & \cdots & & \\ & -\nu & (1+2\nu) & -\nu & \\ & & \cdots & \cdots & \cdots & \\ & & & \cdots & \cdots & \\ & & & -\nu & (1+2\nu) \end{bmatrix} \begin{bmatrix} u_2^{n+1} \\ u_3^{n+1} \\ \cdots \\ \cdots \\ u_i^{n+1} \\ \cdots \\ \cdots \\ u_{I-1}^{n+1} \end{bmatrix} = \begin{bmatrix} u_2^n + \nu u_1^{n+1} \\ u_3^n \\ \cdots \\ \cdots \\ u_i^n \\ \cdots \\ \cdots \\ u_{I-1}^n + \nu u_I^{n+1} \end{bmatrix}. \tag{10.6}$$

For didactic reasons we take u_1^{n+1} and u_I^{n+1} to be known from Dirichlet boundary conditions. Neumann and Cauchy conditions can equally well be applied, but are somewhat more complicated to implement. The solution at time level $n+1$ is determined by solving this system of equations. The implicit method is consequently very computationally demanding compared to the explicit method, but since it is unconditionally stable it is possible to use larger time steps. In the present case, the matrix is tridiagonal, which is advantageous from a computational standpoint, since the problem can be solved using *e.g.* the Thomas algorithm, a simplified version of Gaussian elimination.

10.2 Semi-implicit schemes

Semi-implicit schemes evaluate the spatial derivative at an average of the time levels n and $n + 1$ instead of only at $n + 1$ as in the fully-implicit case. If $F(x, y, t)$ is a term comprising spatial derivatives of a given scalar $T(x, y, t)$, we can consider the general expression for a discretised version of the equation for the time evolution of $u_{i,j}$:

$$\frac{du}{dt} = F(x, y) \; \Rightarrow \; \frac{u_{i,j}^{n+1} - u_{i,j}^{n}}{\Delta t} = (1 - \beta)F_{i,j}^{n} + \beta F_{i,j}^{n+1}, \qquad (10.7)$$

where $\beta = 0$ yields an explicit scheme, $\beta = 1$ a fully implicit scheme and $0 < \beta < 1$ a semi-implicit scheme.

A commonly used semi-implicit method is given by the Crank-Nicolson scheme, in which $\beta = 0.5$ and the time derivative is expressed with the usual Euler-forward scheme. The term comprising spatial derivatives is therefore centred at time level $n + 1/2$, which in fact turns this scheme into a trapezoidal implicit scheme in time. By carrying out a Taylor expansion around $(i, n + 1/2)$, one can verify that this scheme is characterised by a second-order accuracy in time, which represents an appreciable improvement with regard to the first-order accuracy of the Euler-forward explicit scheme.

10.2.1 The one-dimensional (1D) diffusion equation

The heat equation (8.4) is usually associated with centred differencing in space. When using a semi-implicit time scheme it becomes

$$\frac{u_i^{n+1} - u_i^{n}}{\Delta t} = (1 - \beta)A\frac{u_{i+1}^{n} - 2u_i^{n} + u_{i-1}^{n}}{(\Delta x)^2} + \beta A\frac{u_{i+1}^{n+1} - 2u_i^{n+1} + u_{i-1}^{n+1}}{(\Delta x)^2}.$$

$$(10.8)$$

The semi-implicit Crank-Nicolson scheme ($\beta = 0.5$) results in a numerical precision of second order both in time and space and hence the truncation error is of $\mathcal{O}\left[(\Delta t)^2, (\Delta x)^2\right]$.

10.2.2 Two-dimensional (2D) pure gravity waves

Let us now discretise the equations for two-dimensional shallow-water gravity waves, *viz.* Equations (7.25) without the Coriolis terms, using the Crank-Nicolson scheme on a C-grid and an Euler-forward time integration:

$$u_{i,j}^{n+1} = u_{i,j}^n - \frac{g\Delta t}{2\Delta x}\left(h_{i+1,j}^{n+1} - h_{i,j}^{n+1} + h_{i+1,j}^n - h_{i,j}^n\right), \tag{10.9a}$$

$$v_{i,j}^{n+1} = v_{i,j}^n - \frac{g\Delta t}{2\Delta y}\left(h_{i,j+1}^{n+1} - h_{i,j}^{n+1} + h_{i,j+1}^n - h_{i,j}^n\right), \tag{10.9b}$$

$$h_{i,j}^{n+1} = h_{i,j}^n - \tag{10.9c}$$

$$H\Delta t\left(\frac{u_{i,j}^{n+1} - u_{i-1,j}^{n+1} + u_{i,j}^n - u_{i-1,j}^n}{2\Delta x} + \frac{v_{i,j}^{n+1} - v_{i,j-1}^{n+1} + v_{i,j}^n - v_{i,j-1}^n}{2\Delta y}\right).$$

$$\tag{10.9d}$$

A stability analysis of these equations can be undertaken by inserting the wave solutions

$$(u^n, v^n, h^n) = (u_0, v_0, h_0)\lambda^n e^{I(ki\Delta x + lj\Delta y)}, \tag{10.10}$$

and we find

$$u_0\left(1-\lambda\right) = \frac{g\Delta t}{2\Delta x}\left(1+\lambda\right)\left(e^{Ik\Delta x}-1\right)h_0, \tag{10.11a}$$

$$v_0\left(1-\lambda\right) = \frac{g\Delta t}{2\Delta y}\left(1+\lambda\right)\left(e^{Il\Delta y}-1\right)h_0, \tag{10.11b}$$

$$h_0\left(1-\lambda\right) = H\Delta t\left(1+\lambda\right)\left(\frac{1-e^{-Ik\Delta x}}{\Delta x}u_0 + \frac{1-e^{-Il\Delta y}}{\Delta y}v_0\right). \tag{10.11c}$$

By eliminating u_0, v_0 and h_0, we obtain the following quadratic equation for λ:

$$\lambda^2 - 2\lambda\frac{1-B}{1+B} + 1 = 0,$$

$$B \equiv 2gH\Delta t^2\left[\frac{\sin^2\left(k\Delta x/2\right)}{\Delta x^2} + \frac{\sin^2\left(l\Delta y/2\right)}{\Delta y^2}\right], \tag{10.12}$$

with the two roots

$$\lambda_{1,2} = \frac{1-B\pm 2i\sqrt{B}}{1+B}. \tag{10.13}$$

These two amplification factors have the absolute value

$$|\lambda_{1,2}|^2 = \left(\frac{1-B}{1+B}\right)^2 + \left(\frac{2\sqrt{B}}{1+B}\right)^2 = 1,$$

and thus the scheme is unconditionally stable. The scheme is also said to be "neutrally stable" since $|\lambda| = 1$ is just at the edge of stability. This example of an application of the Crank-Nicolson scheme shows the power of semi-implicit methods as these both decrease the temporal truncation error from $O(\Delta t)$ to $O\left[(\Delta t)^2\right]$ and make the scheme unconditionally stable.

To solve the system constituted by Equations (10.9), as will be done in next section, the quantities $\delta_x u^{n+1}$ and $\delta_y v^{n+1}$ can be eliminated from the third of these equations by applying the operators δ_x and δ_y to the first and second, respectively, and substituting the results into the third equation. This yields a set of equations for the height h in the form of a linear matrix system involving each grid point of the domain. This problem can be solved using $e.g.$ an iterative procedure similar to those further discussed in the previous chapter:

1. Make a first guess h^{n+1} which is usually h^n.
2. At each of the grid points the value of h^{n+1} has to satisfy the equation.
3. The preceding step is repeated as many times as needed to make the change at every point less than some pre-assigned small value.

10.3 The semi-implicit method of Kwizak and Robert

When considering the shallow water equations Kwizak and Robert (1971) chose to use the leap-frog scheme and a semi-implicit difference system for variables at time level $n + 1$. The governing equations can be written in a compact form:

$$\frac{\partial u}{\partial t} = -g\frac{\partial h}{\partial x} + A_u, \tag{10.14a}$$

$$\frac{\partial v}{\partial t} = -g\frac{\partial h}{\partial y} + A_v, \tag{10.14b}$$

$$\frac{\partial h}{\partial t} = -H\left(\frac{\partial u}{\partial x} + \frac{\partial v}{\partial y}\right) + A_h, \tag{10.14c}$$

where A_u, A_v and A_h represent terms that were omitted in Equations (7.25) describing the propagation of pure gravity waves. This time we apply implicit differencing over a time interval $2\Delta t$ centred around time n for the terms containing spatial derivatives by using $\beta = 0.5$ with time steps $n-1$ and $n+1$ rather than n and $n+1$ as previously. Second-order centred schemes are used for spatial derivatives and the leap-frog scheme for the time derivative, and hence the discretised system is

$$u^{n+1} = u^{n-1} - g\Delta t\left(\delta_x h^{n-1} + \delta_x h^{n+1}\right) + 2\Delta t A_u^n, \tag{10.15a}$$

$$v^{n+1} = v^{n-1} - g\Delta t\left(\delta_y h^{n-1} + \delta_y h^{n+1}\right) + 2\Delta t A_v^n, \tag{10.15b}$$

$$h^{n+1} = h^{n-1} - H\Delta t\left(\delta_x u^{n-1} + \delta_y v^{n-1} + \delta_x u^{n+1} + \delta_y v^{n+1}\right) + 2\Delta t A_h^n. \tag{10.15c}$$

We now apply the operator δ_x to the first and δ_y to the second of these equations, and add the results. By introducing the notation

$$\delta_{xx} h = \delta_x\left(\delta_x h\right) \quad \text{and} \quad \delta_{yy} h = \delta_y\left(\delta_y h\right), \tag{10.16}$$

we obtain

$$\begin{aligned}
\left(\delta_x u + \delta_y v\right)^{n+1} = &\left(\delta_x u + \delta_y v\right)^{n-1} \\
&- g\Delta t\left[\left(\delta_{xx} + \delta_{yy}\right) h^{n-1} + \left(\delta_{xx} + \delta_{yy}\right) h^{n+1}\right] \\
&+ 2\Delta t\left(\delta_x A_u + \delta_y A_v\right)^n.
\end{aligned}$$

Substituting the right-hand side into Equation (10.15c), and defining the finite-difference Laplacian by

$$\nabla_*^2 \equiv \delta_{xx} + \delta_{yy}, \tag{10.17}$$

we find that

$$\begin{aligned}
h^{n+1} = &\ h^{n-1} - 2H\Delta t\left(\delta_x u + \delta_y v\right)^{n-1} + gH\Delta t^2\left(\nabla_*^2 h^{n-1} + \nabla_*^2 h^{n+1}\right) \\
&+ 2\Delta t\left[A_h - H\Delta t\left(\delta_x A_u + \delta_y A_v\right)\right]^n. \tag{10.18}
\end{aligned}$$

By, in addition, introducing the definitions

$$F^{n-1} \equiv h^{n-1} - 2H\Delta t\left(\delta_x u + \delta_y v\right)^{n-1} + gH\Delta t^2 \nabla_*^2 h^{n-1}, \tag{10.19}$$

$$G^n \equiv 2\Delta t\left[A_h - H\Delta t\left(\delta_x A_u + \delta_y A_v\right)\right]^n, \tag{10.20}$$

this result can be formulated as

$$h^{n+1} - gH\Delta t^2 \nabla_*^2 h^{n+1} = F^{n-1} + G^n, \tag{10.21}$$

where the terms have been arranged to show that at time level n, the right-hand side of this equation is known at all grid points. This is an elliptic PDE reminiscent of the *Helmholtz* equation

$$\nabla^2 h + ah + b\left(x, y\right) = 0. \tag{10.22}$$

Several methods are available for resolving this standard problem. Once it has been solved for h^{n+1}, then u^{n+1} and v^{n+1} can be obtained directly from Equations (10.15).

11. The Semi-Lagrangian Technique

In an Eulerian advection scheme an observer at a fixed point in point in space watches the surroundings. Such schemes work well on structured grids of the type to be discussed in Chapter 12, but often lead to unnecessarily restrictive time steps imposed by the requirement of computational stability. In a Lagrangian advection scheme the observer watches the ambient world evolve while travelling "on board" a fluid particle. An advantage with a Lagrangian scheme is that one can use much larger time steps than in the Eulerian case. A disadvantage is, however, that a regularly spaced set of particles will in most cases subsequently evolve into one which is highly irregularly spaced and important characteristics of the flow may consequently be lost. The advantage with the semi-Lagrangian advection schemes is that they combine the regular resolution of the Eulerian schemes with the enhanced stability of the Lagrangian ones. Robert (1981) proposed using a semi-Lagrangian scheme for the treatment of the advective part of the equations (see *e.g.* Staniforth and Côté (1991) for a general review).

11.1 The 1D linear advection equation

To present the basic idea underlying the semi-Lagrangian method in its simplest context let us first examine the one-dimensional advection equation formulated within an Eulerian framework:

$$\frac{\partial F}{\partial t} + c\frac{\partial F}{\partial x} = 0, \tag{11.1}$$

where c is a given constant velocity and F the passively advected tracer such as *e.g.* the temperature or moisture. The most straightforward

---

How to cite this book chapter:
Döös, K., Lundberg, P., and Campino, A. A. 2022. *Basic Numerical Methods in Meteorology and Oceanography*, pp. 109–116. Stockholm: Stockholm University Press. DOI: https://doi.org/10.16993/bbs.k. License: CC BY 4.0

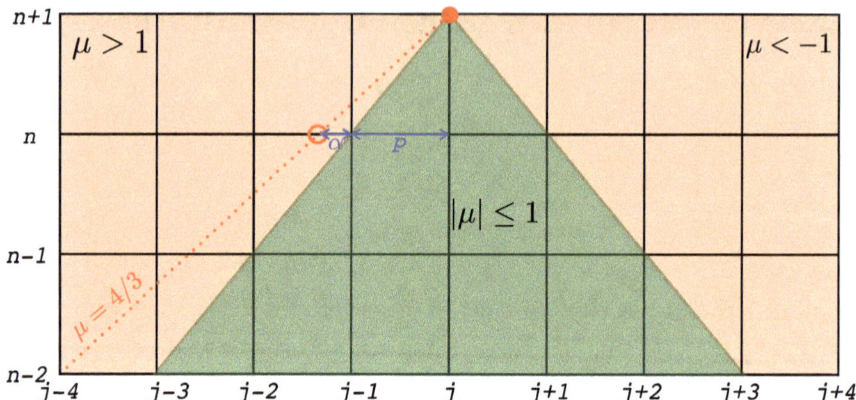

Figure 11.1. The green stable region for the advection equation discretised with centred finite differences in both time and space. The maximum distance you may travel during one time step Δt is one grid length Δx. The red dotted line illustrates how advection with a speed $c = 4\Delta x/3\Delta t$ takes place in the red unstable region, which can, however, be solved numerically with a semi-Lagrangian scheme.

discretisation of this equation, as presented in Chapter 4, is centred differencing in both time and space:

$$\frac{F_j^{n+1} - F_j^{n-1}}{2\Delta t} + c\frac{F_{j+1}^n - F_{j-1}^n}{2\Delta x} = 0. \tag{11.2}$$

This scheme will only yield a stable solution when integrated with a Courant number $\mu \leq 1$ as shown by the green area in Figure 11.1.

Equation (11.1) can be formulated in a Lagrangian framework instead of an Eulerian one, resulting in

$$\frac{DF}{Dt} = 0, \tag{11.3}$$

where the Lagrangian derivative is defined as $D/Dt \equiv \partial/\partial t + c\partial/\partial x$. Equation (11.3) simply shows how the value of F is constant along the corresponding trajectory.

In discretised space this implies that F_j^{n+1} must be equal to the value of F at time step n, which can be expressed as

$$F_j^{n+1} = F_*^n, \tag{11.4}$$

where $*$ symbolises the spatial position at time level n, which is normally not a grid point and in our case is where the red dotted line in Figure 11.1 crosses time level n (shown as the red circle located between $j-2$ and $j-1$). The value of F_*^n can hence be obtained by interpolation between these grid points:

$$F_*^n = \alpha F_{j-2}^n + (1 - \alpha)\, F_{j-1}^n. \tag{11.5}$$

Here $\alpha = \mathrm{frac}\,(\mu)$, where $0 \le \alpha < 1$, is the fractional part of the Courant number. Making use of the integer part of the Courant number, $p \equiv \mathrm{int}\,(\mu)$, $viz.$ $\mu = p + \alpha$, this relationship can be expressed in a more general way as

$$F_*^n = \alpha F_{j-p-1}^n + (1 - \alpha)\, F_{j-p}^n. \tag{11.6}$$

In our example $p = 1$ and $\alpha = 1/3$, as shown in blue in Figure 11.1. The discrete expression can be obtained from Equations (11.4) and (11.6), resulting in

$$F_j^{n+1} = \alpha F_{j-p-1}^n + (1 - \alpha)\, F_{j-p}^n. \tag{11.7}$$

Figure 11.2 shows how this semi-Lagrangian discretisation of the linear 1D advection equation can be used with $\mu = 4/3$ (left panel) but is clearly unstable (right panel) for the centred scheme in time and space, which was presented in Chapter 4 and Figure 4.4.

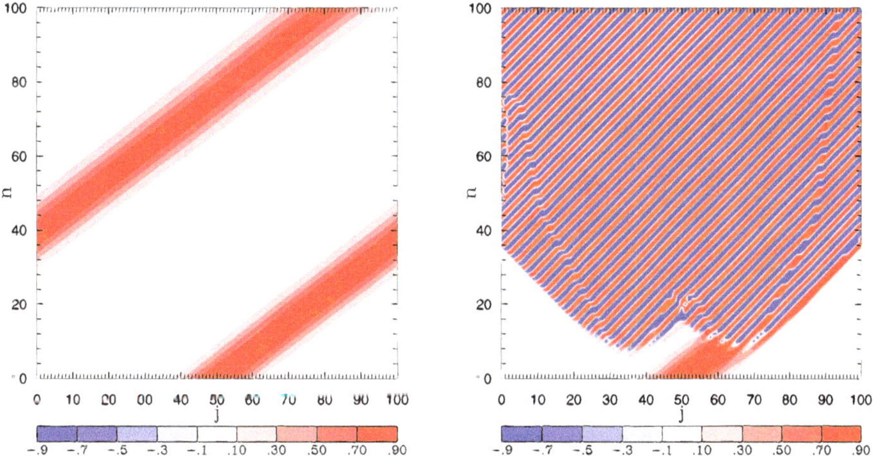

Figure 11.2. Hovmöller diagrams of the linear advection equation integrated numerically with $\mu = 4/3$. A semi-Lagrangian scheme has been used for the left panel and a centred scheme in both time and space for the right panel. Note that the integration with the centred scheme blows up after just a few time steps, cf. Figure 4.4.

11.2 Stability analysis

A von Neumann stability analysis is undertaken of the semi-Lagrangian discretisation of the 1D linear advection equation discussed above. We search for a solution of the form $F_j^n = \lambda^n F_0 e^{ikj\Delta x}$ that we substitute in Equation (11.7), which results in the amplification factor

$$\lambda = \alpha e^{-ik(p+1)\Delta x} + (1-\alpha)e^{-ikp\Delta x} = e^{-ikp\Delta x}\left(1 - \alpha + \alpha e^{-ik\Delta x}\right). \quad (11.8)$$

The scheme is stable for $|\lambda| \leq 1$, which is why we consider

$$|\lambda|^2 = \left|e^{-ikp\Delta x}\right|^2 \left|\left(1 - \alpha + \alpha e^{-ik\Delta x}\right)\right|^2. \quad (11.9)$$

We use Euler's formula, $e^{i\beta} = \cos\beta + i\sin\beta$ and the associated absolute-value result $|\cos\beta + i\sin\beta| = 1$, which leads to

$$|\lambda|^2 = 1 - 2\alpha(1 - \alpha)\left[1 - \cos(k\Delta x)\right]. \quad (11.10)$$

The minimum value of this expression is obtained when $\cos(k\Delta x) = -1$, which yields

$$|\lambda|^2 = (1 - 2\alpha)^2 \leq 1. \quad (11.11)$$

The maximum is obtained for $\cos(k\Delta x) = 1$:

$$|\lambda|^2 = 1. \quad (11.12)$$

The scheme is hence unconditionally stable. The time step can thus be much larger than in the case of *e.g.* the leap-frog scheme. A larger time step will, however, decrease the numerical accuracy in a GCM, a fact that must be taken into account. In a GCM employed for numerical weather prediction this will nevertheless make it possible to use a time step approximately six times larger than in the leap-frog case. The absolute stability of the semi-Lagrangian scheme can be understood in the sense that taking one single step along the flow in both time and space is a way to adjust the spatial resolution. This is like prescribing the spatial resolution coarser when possible, "as if the Courant number had been equal to one".

11.3 The advection equation with variable velocity

The derivations in the previous section can be extended to the 1D non-linear advection equation for a tracer, where we replace the constant phase speed c with a velocity, which varies in time and space.

$$\frac{DF}{Dt} = \frac{\partial F}{\partial t} + u\frac{\partial F}{\partial x} = 0, \quad (11.13)$$

where $u = u(x,t)$ is a given velocity and F the passively advected tracer such as *e.g.* the temperature or moisture. It can also, as in the momentum equations, be the velocity itself so that $F(x,t) = u(x,t)$ We consider the case where the velocity u_j^n is known at all grid points in space and time. The centred scheme in both time and space in the Eulerian framework is then

$$\frac{F_j^{n+1} - F_j^{n-1}}{2\Delta t} + u_j^n \frac{F_{j+1}^n - F_{j-1}^n}{2\Delta x} = 0. \tag{11.14}$$

The Courant number is now $\mu = u_j^n \Delta t / \Delta x$ and hence variable in time and space but should never exceed 1 anywhere and anytime for this centred case. Figure 11.3 shows an example of this, where the velocity has been prescribed as varying in both time and space, and where the maximum velocity was just at the limit of stability ($\mu = 1$).

The semi-Lagrangian scheme for a variable prescribed velocity is, just like in the linear case, based on that one can compute the value F_j^{n+1} by following a trajectory backwards from its upwind position at a previous time step. The "exact" backward trajectory from the point where F_j^{n+1} is illustrated by the blue curve AC in Figure 11.4. For our calculations it will be approximated by the red straight line $A'C$. Equation (11.13) states that the scalar F remains constant along a fluid path or trajectory. The integration along the approximated trajectory of Equation (11.13) is thus

$$F_j^{n+1} = F_*^{n-1}, \tag{11.15}$$

where F_*^{n-1} is located at A', which subsequently is to be determined. The particle displacement in the x-direction over the two time steps from point A' to C is $2u_*^n \Delta t$. Here u_*^n is the interpolated velocity at point B:

$$u_*^n = \alpha u_{j-p-1}^n + (1 - \alpha)\, u_{j-p}^n, \tag{11.16}$$

where $p = \text{int}(\mu)$ and $\alpha = \mu - p$. This is the same interpolation as for the case with a constant velocity but with the difference that the Courant number now depends on a variable velocity u_*^n:

$$\mu = \frac{u_*^n \Delta t}{\Delta x}. \tag{11.17}$$

Since both α and u_*^n depend on each other we need to iterate in order find the point B with the corresponding velocity u_*^n. Once this is found

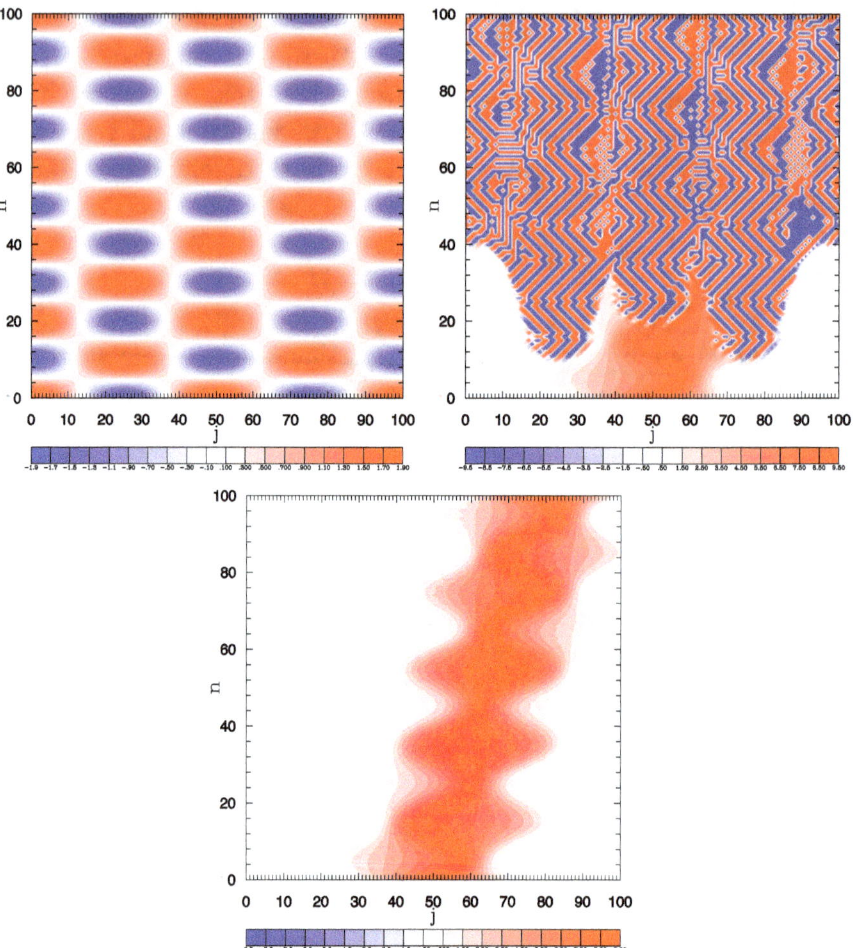

Figure 11.3. Hovmöller diagrams. The prescribed time- and space-dependent velocity u_j^n (upper left panel). The non-linear advection equation integrated numerically with a leap-frog scheme (upper right panel) and with the semi-Lagrangian scheme (lower panel). The Courant number varies over the interval $-1.75 \leq \mu \leq 2.25$, which results in a clearly unstable solution in the leap-frog case but a nice and smooth solution using the semi-Lagrangian framework.

we can determine the point A' and compute the interpolated value of F in this position:

$$F_*^{n-1} = \beta F_{j-q-1}^{n-1} + (1 - \beta) F_{j-q}^{n-1}, \qquad (11.18)$$

where $q \equiv \mathrm{int}(2\mu)$ and $\beta = 2\mu - q$. Here the factor "2" is due to the two time steps separating F_*^{n-1} from F_j^{n+1}.

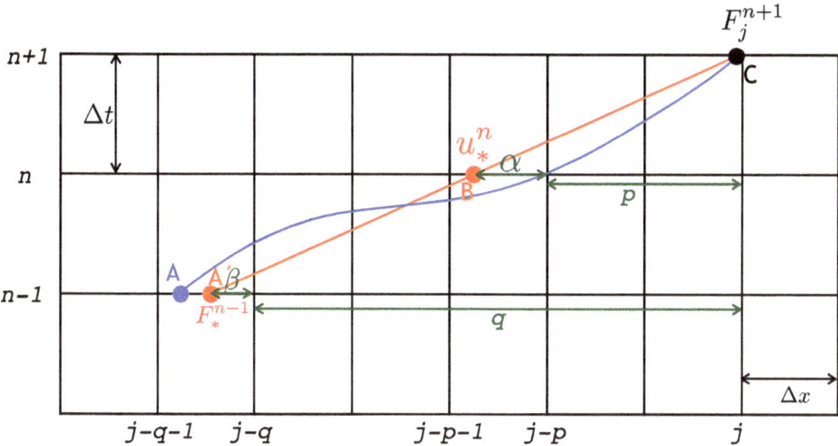

Figure 11.4. Schematic representing the semi-Lagrangian scheme with variable-velocity advection. The "real" (blue curve) and approximate (red line) trajectories that arrive at grid point j at time level $n+1$. Here α and p are the interpolation coefficients used to calculate u_*^n, β and q, the quantities used for the interpolation of F_*^{n-1}.

The semi-Lagrangian calculation will thus go through the following stages:

1. A first guess of u_*^n, which could be $u_*^n = u_j^n$.
2. Compute μ, α and p with this u_*.
3. Compute a new $u^* = \alpha u_{j-p-1}^n + (1 - \alpha)\, u_{j-p}^n$.
4. Iterate over stages 2 and 3 until no significant improvement is obtained.
5. Compute $q \equiv \mathrm{int}(2\mu)$ and $\beta = 2\mu - q$.
6. Finally use $F_j^{n+1} = F_*^{n-1} = \beta F_{j-q-1}^{n-1} + (1 - \beta)\, F_{j-q}^{n-1}$.

The semi-Lagrangian formulation above is only valid for one-dimensional problems on a regular grids but can easily be extended to curvilinear grids as well as two and three-dimensional flows. The interpolations presented here have all been linear for didactic reasons. The disadvantage is, however, that the linear interpolation creates too much diffusion of the tracers. The semi-Lagrangian formulations, which are employed in circulation models today, for this reason use higher-order schemes such as cubic interpolations. The clear advantage of the semi-Lagrangian scheme over the Eulerian ones is that one can use much larger time steps Δt. The accuracy of the semi-Lagrangian scheme discusssed in the present chapter is only of first order and will decrease with longer time steps. The time step Δt is often chosen to be 5 to 10 times larger for a semi-Lagrangian scheme than for an Eulerian one. A drawback with the semi-Lagrangian schemes has been that they were not as mass conserving as the Eulerian ones, which has been rectified in the more recent formulations. When applying the semi-Lagrangian method to the shallow water equations with forcing, dissipation and bottom topography some additional difficulties will arise.

12. Model Coordinates

In order to present some 3D modelling in the next chapter we will here show the different types of vertical coordinates that are used for oceanic and atmospheric circulation models.

12.1 Oceanic vertical coordinates

The most common vertical coordinate systems used in ocean circulation models are presented in Figure 12.3. They are of basically three types: z-coordinates (constant depth), terrain-following sigma coordinates and isopycnic coordinates with density layers. The first oceanic general circulation model, developed by Bryan and Cox (1967), used fixed z-coordinates with a rigid lid as illustrated by the top right panel of Figure 12.1. The rigid-lid approximation was replaced (Killworth et al., 1991) by treating the fast barotropic mode separately, *viz.* introducing a free surface. Another improvement of the fixed z-coordinates was achieved by Pacanowski and Gnanadesikan (1998) by adjusting the thickness of the deepest layer in order to match the total depth as illustrated by the middle left panel of Figure 12.3. Adcroft and Campin (2004) introduced a time dependence in what is known as z^*-coordinates, where all layers were adjusted to the free-surface elevation variations like an accordion (middle right panel of Figure 12.3). Terrain-following vertical coordinates were suggested by Phillips (1957) for atmospheric forecasting models

12.1.1 Fixed-depth coordinates

A simple example of a vertical discretisation is the one of the continuity equation, which is used in many Ocean General Circulation Models

How to cite this book chapter:
Döös, K., Lundberg, P., and Campino, A. A. 2022. *Basic Numerical Methods in Meteorology and Oceanography*, pp. 117–134. Stockholm: Stockholm University Press. DOI: https://doi.org/10.16993/bbs.l. License: CC BY 4.0

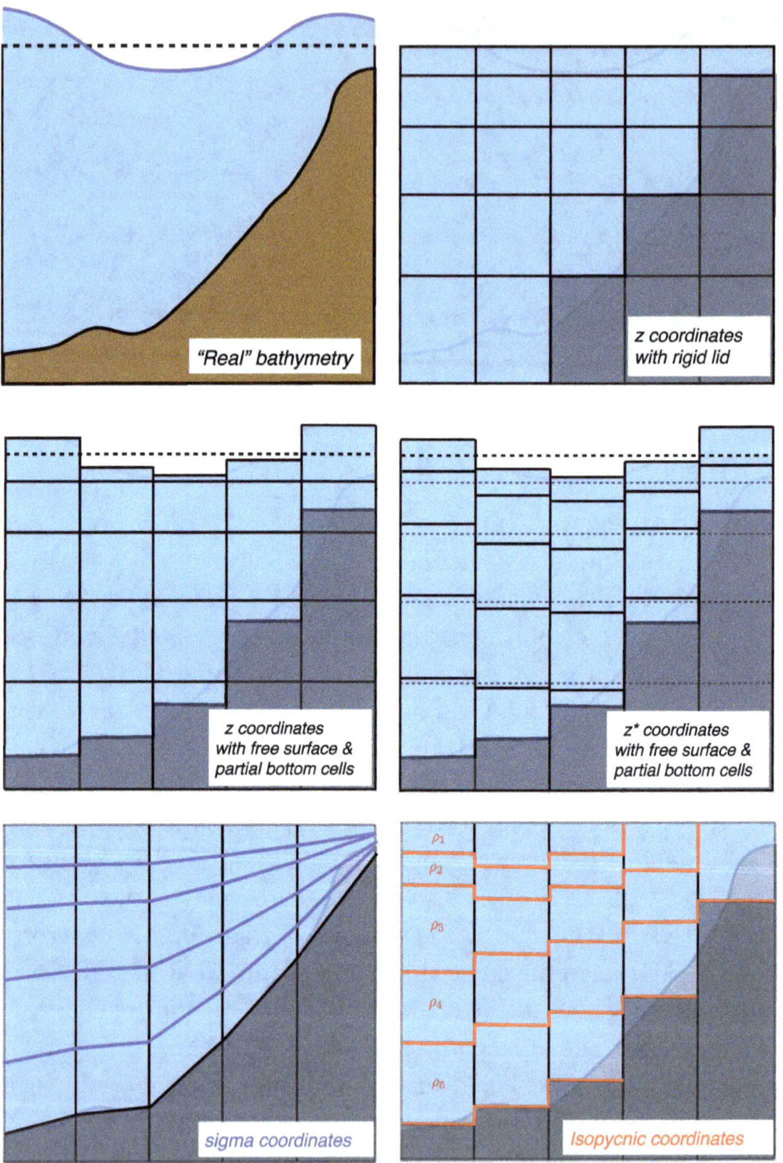

Figure 12.1. Different ocean-model vertical coordinates.

(OGCMs), based on the B-grid with fixed-depth levels as shown in Figure 12.2. The continuity equation

$$\frac{\partial u}{\partial x} + \frac{\partial v}{\partial y} + \frac{\partial w}{\partial z} = 0 \qquad (12.1)$$

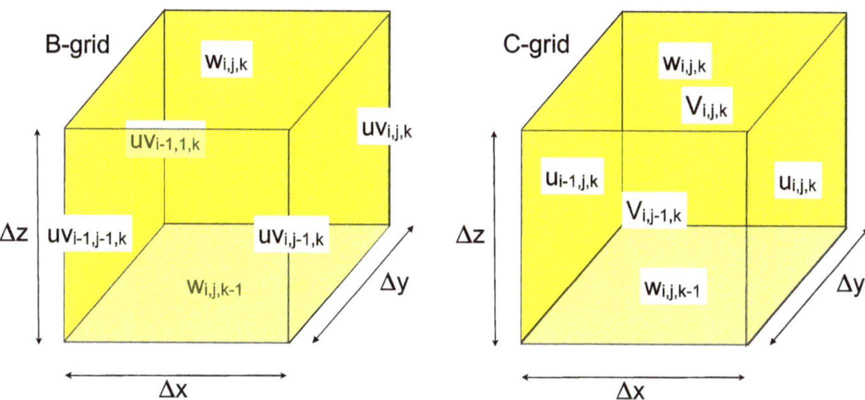

Figure 12.2. Finite-difference boxes for the B-grid and the C-grid with fixed-depth level coordinates.

can be discretised on a B-grid as

$$w_{i,j,k} = w_{i,j,k-1} - \Delta z \left[\frac{(u_{i,j,k} + u_{i,j-1,k}) - (u_{i-1,j,k} + u_{i-1,j-1,k})}{2\Delta x} \right.$$

$$\left. + \frac{(v_{i,j,k} + v_{i-1,j,k}) - (v_{i,j-1,k} + v_{i-1,j-1,k})}{2\Delta y} \right],$$

$$(12.2)$$

or on a C-grid as

$$w_{i,j,k} = w_{i,j,k-1} - \Delta z \left(\frac{u_{i,j,k} - u_{i-1,j,k}}{\Delta x} + \frac{v_{i,j,k} - v_{i,j-1,k}}{\Delta y} \right). \qquad (12.3)$$

Equation (12.2) is integrated from the bottom upwards with the boundary condition $w_{i,j,k=0} = 0$ at the bottom of the ocean. The interpretation of this equation is that the sum of all the volume fluxes in or out of the grid box is zero. An alternative way to derive the vertical velocity is therefore to consider that the sum of the volume transport through the six grid-box walls must be zero due to the incompressibility. This sum for the C-grid box is hence

$$(u_{i,j,k} - u_{i-1,j,k}) \Delta y \Delta z + (v_{i,j,k} - v_{i,j-1,k}) \Delta x \Delta z$$

$$+ (w_{i,j,k} - w_{i,j,k-1}) \Delta x \Delta y = 0, \qquad (12.4)$$

which becomes identical to Equation (12.3) by a reformulation. Note that we have for simplicity used a k that decreases with depth in order to have the upward positive direction as k increases. In most OGCMs, however, k will decrease with depth, with the surface layer counting as $k = 1$.

12.1.2 Variable-depth coordinates

The layer thickness is in most of today's OGCMs a function of both space and time as shown in Figure 12.1. On a C-grid the mass transports through the eastern, northern and upper faces, respectively, of the i, j, k grid box at time step n are given by

$$U_{i,j,k}^n = \rho_{i,j,k}^n u_{i,j,k}^n \Delta y_{i,j} \Delta z_{i,j,k}^n, \tag{12.5}$$

$$V_{i,j,k}^n = \rho_{i,j,k}^n v_{i,j,k}^n \Delta x_{i,j} \Delta z_{i,j,k}^n, \tag{12.6}$$

$$W_{i,j,k}^n = \rho_{i,j,k}^n w_{i,j,k}^n \Delta x_{i,j} \Delta y_{i,j}, \tag{12.7}$$

where the unit is kg/s. The continuity equation, which expresses conservation of mass, states that

$$\frac{\partial \rho}{\partial t} + \frac{\partial(\rho u)}{\partial x} + \frac{\partial(\rho v)}{\partial y} + \frac{\partial(\rho w)}{\partial z} = 0. \tag{12.8}$$

Integrating Equation (12.8) over a finite grid box of volume $\Delta x \Delta y \Delta z$ we obtain

$$\frac{\partial M_{i,j,k}}{\partial t} + U_{i,j,k} - U_{i-1,j,k} + V_{i,j,k} - V_{i,j-1,k} + W_{i,j,k} - W_{i,j,k-1} = 0, \tag{12.9}$$

where $M_{i,j,k}$ is the mass of the grid box. The rate of mass change of the grid box $\partial M_{i,j,k}/\partial t$ can be either due to compression in a compressible GCM or to grid-box volume change, which in a GCM is generally due to the time dependence of the vertical resolution so that the thicknesses of model layers vary in time.

The mass of the grid box is

$$M_{i,j,k}^n = \rho_{i,j,k}^n \Delta x_{i,j} \Delta y_{i,j} \Delta z_{i,j,k}^n. \tag{12.10}$$

The vertical mass transport through the top of the grid box is obtained by discretising Equation (12.9) between two stored time levels:

$$W_{i,j,k}^n = W_{i,j,k-1}^n - \left(U_{i,j,k}^n - U_{i-1,j,k}^n + V_{i,j,k}^n - V_{i,j-1,k}^n \right.$$
$$\left. + \frac{\rho_{i,j,k}^{n+1} \Delta z_{i,j,k}^{n+1} - \rho_{i,j,k}^{n-1} \Delta z_{i,j,k}^{n-1}}{2\Delta t} \Delta x_{i,j} \Delta y_{i,j} \right), \tag{12.11}$$

which is computed by integration from the bottom and upwards with the bottom boundary condition $W_{i,j,0} = 0$. The vertical velocity w can then be deduced from Equation 12.7.

In many OGCMs, the fluid is considered to be incompressible and thus the density is taken to be constant in the equation above. The vertical volume transport through the top of the grid box then becomes

$$
W_{i,j,k}^{n} = W_{i,j,k-1}^{n} - \left(U_{i,j,k}^{n} - U_{i-1,j,k}^{n} + V_{i,j,k}^{n} - V_{i,j-1,k}^{n} \right.
$$
$$
\left. + \frac{\Delta z_{i,j,k}^{n+1} - \Delta z_{i,j,k}^{n-1}}{2\Delta t} \Delta x_{i,j} \Delta y_{i,j} \right),
$$

(12.12)

where $\mathcal{U}, \mathcal{V}$ and $\mathcal{W}$ are the volume transports in the unit m^3/s.

12.2 Atmospheric vertical coordinates

Instead of depth/height as vertical coordinate in our system of equations it is possible to use other quantities. The density varies with latitude and height/depth which makes the equations sometimes less easy to use than an alternative system which employs other quantities, such as pressure, sigma or potential temperature for the atmosphere and density or sigma for the ocean, as the vertical coordinate. These coordinates may facilitate solving the complete equations of motion.

12.2.1 Generalised vertical coordinates

We can derive a system of equations for a generalised vertical coordinate ζ, which is assumed to be related to the height/depth by a single-valued monotonic function. When we transform the vertical coordinate a variable $u(x, y, z, t)$ becomes $a(x, y, \zeta(x, y, z, t), t)$. The horizontal coordinates remain the same. Let s represent x, y or t. From Figure 12.3 we see that

$$
\frac{C - A}{\Delta s} = \frac{B - A}{\Delta s} + \frac{C - B}{\Delta z} \frac{\Delta z}{\Delta s},
$$

(12.13)

so that

$$
\left(\frac{\partial a}{\partial s} \right)_{\zeta} = \left(\frac{\partial a}{\partial s} \right)_{z} + \left(\frac{\partial a}{\partial z} \right)_{s} \left(\frac{\partial z}{\partial s} \right)_{\zeta},
$$

(12.14)

where

$$
\frac{\partial a}{\partial \zeta} = \frac{\partial a}{\partial z} \frac{\partial z}{\partial \zeta},
$$

(12.15)

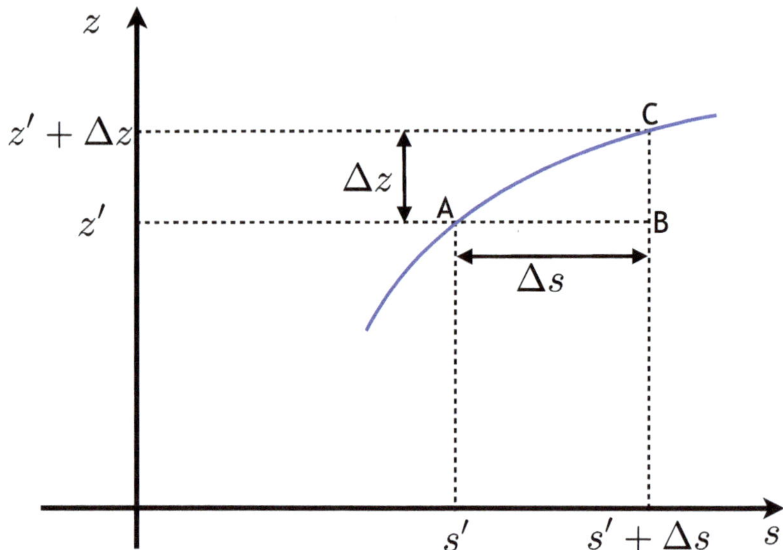

Figure 12.3. Schematic showing the vertical coordinate transformation.

or

$$\frac{\partial a}{\partial z} = \frac{\partial a}{\partial \zeta}\frac{\partial \zeta}{\partial z}. \tag{12.16}$$

Substituting Equation (12.16) in Equation (12.14), we obtain

$$\left(\frac{\partial a}{\partial s}\right)_\zeta = \left(\frac{\partial a}{\partial s}\right)_z + \frac{\partial a}{\partial \zeta}\frac{\partial \zeta}{\partial z}\left(\frac{\partial z}{\partial s}\right)_\zeta. \tag{12.17}$$

From this relationship, we can obtain an equation for a horizontal gradient of the scalar a in ζ coordinates:

$$\nabla_\zeta a = \nabla_z a + \frac{\partial a}{\partial \zeta}\frac{\partial \zeta}{\partial z}\nabla_\zeta z \tag{12.18}$$

and for the horizontal divergence of a vector $\vec{V}$:

$$\nabla_\zeta \cdot \vec{V} = \nabla_z \cdot \vec{V} + \frac{\partial \vec{V}}{\partial \zeta}\cdot\frac{\partial \zeta}{\partial z}\nabla_\zeta z. \tag{12.19}$$

The total derivative of $a(x, y, \zeta, t)$ becomes

$$\frac{Da}{Dt} = \left(\frac{\partial a}{\partial t}\right)_\zeta + \vec{V}\cdot\nabla_\zeta a + \dot{\zeta}\frac{\partial a}{\partial \zeta}. \tag{12.20}$$

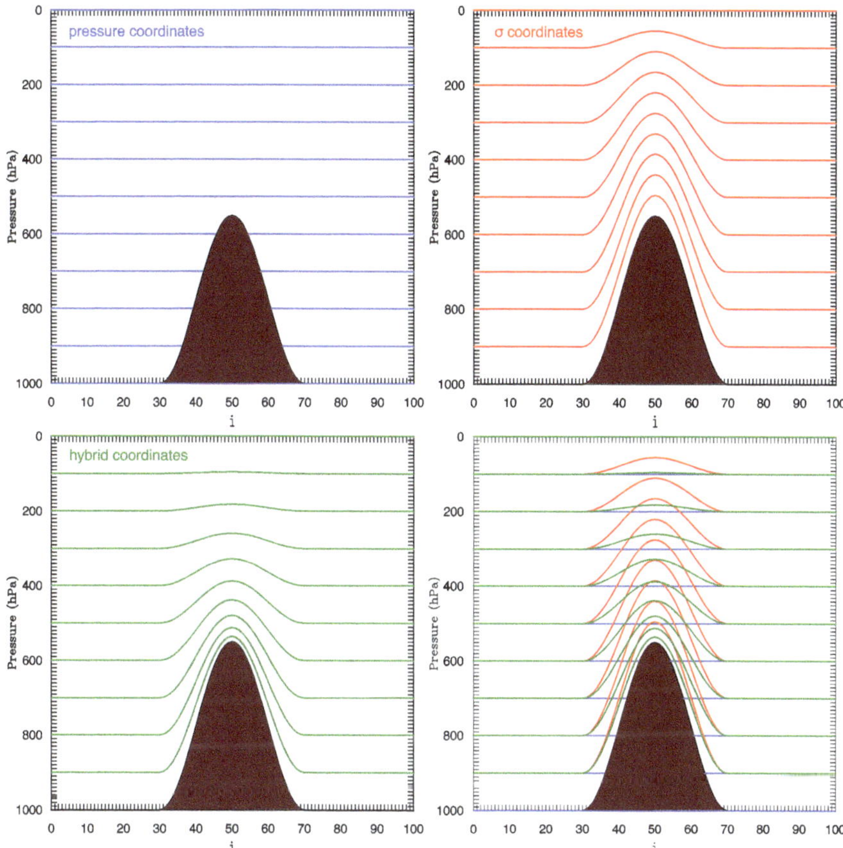

Figure 12.4. Illustration of pressure, sigma and hybrid vertical coordinates in an atmospheric model including terrain. The three types of coordinates are superimposed on each other in the lower right panel.

The three basic (pressure, sigma and hybrid) atmospheric vertical model coordinates will now be examined and are shown in Figure 12.4.

12.2.2 Pressure coordinates

Pressure or isobaric coordinates can be used in the atmosphere when the hydrostatic approximation (cf. Section 13.1) is applied. In pressure coordinates, where $\partial p/\partial\zeta \equiv 1$, the total derivative, Equation (12.20), is given by

$$\frac{da}{dt} = \frac{\partial a}{\partial t} + \vec{V}\cdot\nabla a + \omega\frac{\partial a}{\partial p}, \tag{12.21}$$

where the vertical velocity in pressure coordinates is $\omega \equiv dp/dt$. The continuity equation (12.1) can now be written as

$$\nabla_p \cdot \vec{V} + \frac{\partial \omega}{\partial p} = 0. \tag{12.22}$$

From this expression ω can be computed by integrating from the top of the atmosphere, where $\omega = 0$, downwards over the pressure layers k, which on a C-grid yields

$$\omega_{i,j,k} = \omega_{i,j,k-1} - \Delta p_k \left(\frac{u_{i,j,k} - u_{i-1,j,k}}{\Delta x} + \frac{v_{i,j,k} - v_{i,j-1,k}}{\Delta y} \right). \tag{12.23}$$

This is known as the "the kinematic method" to compute the vertical motion in the atmosphere. This is basically the same method as is used in ocean models expressed by Equation (12.3) and has the clear advantage of being mass conserving. The drawback is, however, that this motion tends to be very "noisy" since it depends on the divergence of the wind and hence on its weak ageostrophic component. In particular, this becomes a problem when the horizontal velocity field is based on observations, as is the case in numerical weather predictions. The "adiabatic method" may then be employed instead by using the thermodynamic energy equation based on the the first law of thermodynamics, see *e.g.* Holton and Hakim (2013). This law, if expressed in the isobaric system in the absence of diabatic heating or cooling, is

$$\frac{\partial T}{\partial t} + u\frac{\partial T}{\partial x} + v\frac{\partial T}{\partial y} + \omega\frac{\partial T}{\partial p} - \frac{\omega}{\rho C_p} = 0. \tag{12.24}$$

By defining the the static stability parameter for the isobaric system as

$$\sigma \equiv \frac{1}{\rho C_p} - \frac{\partial T}{\partial p}, \tag{12.25}$$

where C_p is the specific heat at constant pressure, we can deduce an expression for the vertical motion:

$$\omega_{i,j,k} = \frac{1}{\sigma} \left(\frac{\partial T}{\partial t} + u\frac{\partial T}{\partial x} + v\frac{\partial T}{\partial y} \right). \tag{12.26}$$

There are other methods to calculate the vertical atmospheric motion such as the "vorticity method" (Sawyer, 1949), which as the names indicates, is based on a vorticity equation. Most atmospheric GCMs today, however, use terrain-following vertical coordinates, described below.

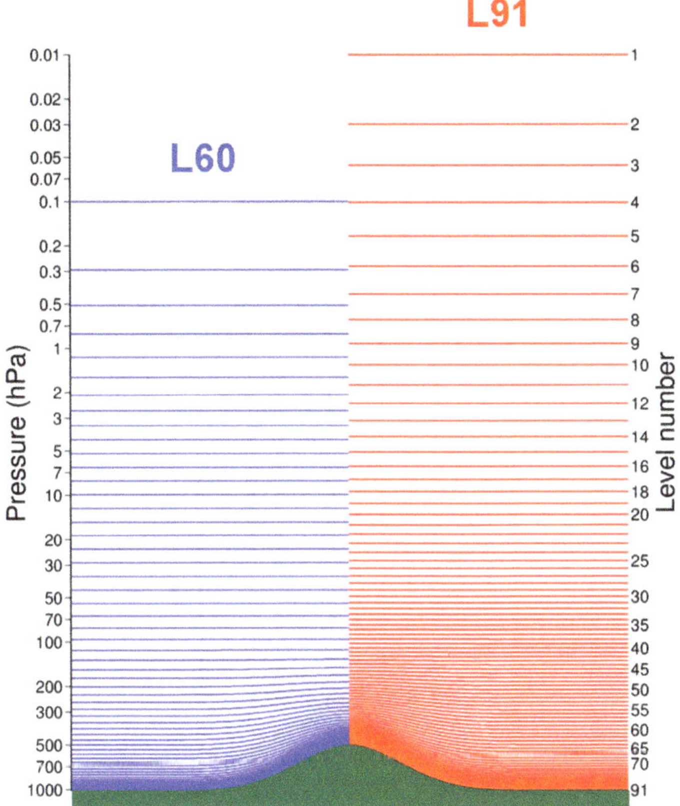

Figure 12.5. Two different vertical resolutions of the hybrid-coordinate model used at the European Centre for Medium-Range Weather Forecasts (ECMWF).

12.2.3 Atmospheric sigma coordinates

The sigma-coordinate system defines the origin at the ground level of the model. The surfaces in the sigma-coordinate system follow the model terrain and are steeply sloped in the regions where terrain itself is strongly inclined. The sigma-coordinate system defines the vertical position in the atmosphere as a ratio of the pressure difference between the location in question and the top of the domain to that of the pressure difference between the origin and the top of the domain. Because it is pressure-based and normalised, it is mathematically easy to cast the governing equations of the atmosphere in a relatively simple form. The sigma coordinate is hence $\sigma = p/p_S$, where $p_S(x, y, z, t)$ is the pressure at the surface of the Earth. The boundary values are consequently $\sigma = 0$ at the top of the atmosphere where $p = 0$ and $\sigma = 1$ at the surface of the Earth. An illustration of the difference between pressure and sigma coordinates is presented in the upper right panel of Figure 12.4.

12.2.4 Hybrid coordinates

The hybrid coordinate system has the properties of sigma coordinates in the lower atmosphere and those of pressure coordinates in the stratosphere. Following Simmons and Burridge (1981) the atmosphere is divided into N_{lev} layers, which are defined by the pressures at the interfaces between them, these pressures given by

$$p_{k+1/2} = a_{k+1/2} + b_{k+1/2}\, p_S \qquad (12.27)$$

for $k = 0, 1, .., N_{lev}$, with $k = 0$ at the top of the atmosphere and $k = N_{lev}$ at the surface of the Earth. The $a_{k+1/2}$ and $b_{k+1/2}$ are constants, the values of which effectively define the vertical coordinate and p_S is the surface pressure. The values a_k and b_k for a 60-layer model are shown in Figure 12.6.

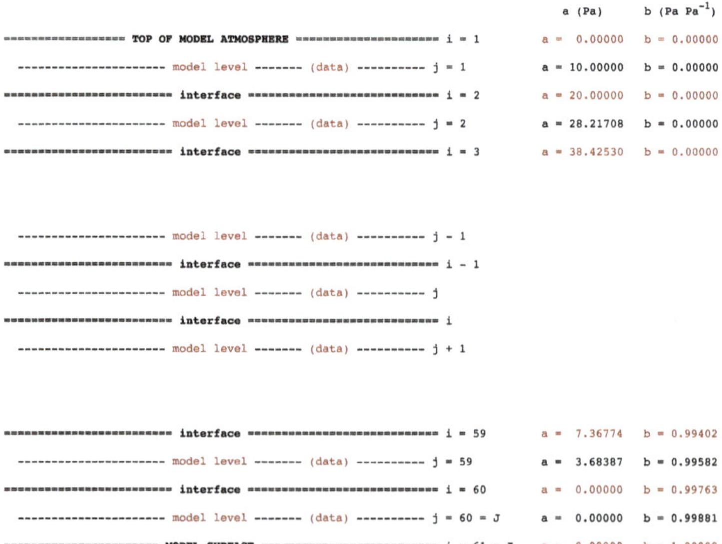

Figure 12.6. Model levels for the 60 layers shown in Figure 12.5, with corresponding a_k and b_k. The middle of each layer in red and the interfaces in black.

The dependent variables, *viz.* the zonal wind u, the meridional wind v, the temperature T and the specific humidity q, are defined in the middle of the layers, where the pressure is defined as

$$p_k = \frac{1}{2}(p_{k-1/2} + p_{k+1/2}) \tag{12.28}$$

for $k = 1, 2, .., N_{lev}$. The vertical coordinate is $\eta = \eta(p, p_S)$ and has the boundary value $\eta(0, p_S) = 0$ at the top of the atmosphere and $\eta(p_S, p_S) = 1$ at the surface of the Earth. Two different vertical resolutions with hybrid coordinates are shown in Figure 12.5.

The vertical mass transport between the layers in a hydrostatic AGCM can be deduced from Equation (12.11) so that

$$W_{i,j,k}^n = W_{i,j,k-1}^n - \left(U_{i,j,k}^n - U_{i-1,j,k}^n + V_{i,j,k}^n - V_{i,j-1,k}^n \right.$$
$$\left. + \frac{\Delta p_{i,j,k}^{n+1} - \Delta p_{i,j,k}^{n-1}}{2g\Delta t} \Delta x_{i,j} \Delta y_{i,j} \right), \tag{12.29}$$

where the hydrostatic approximation has been used.

12.2.5 Isentropic coordinates

Isentropic vertical coordinates use the potential temperature, which is defined as

$$\theta \equiv T \left(\frac{p_0}{p} \right)^{(R_g/C_p)}, \tag{12.30}$$

where R_g is the specific gas constant, C_p the specific heat at constant pressure and p_0 the reference pressure, where $\theta = T$.

Isentropic vertical coordinates are convenient when the motion is adiabatic since the potential temperature θ of an air parcel is then conserved. In the absence of diabatic processes and mixing, air flows along the θ surfaces, which then act as "material" surfaces. A Lagrangian or quasi-Lagrangian vertical coordinate is one that moves with the fluid. The isentropic coordinate system therefore qualifies as such. This is its main advantage since that "vertical" motion is very weak if the flow is quasi-adiabatic, which reduces finite-difference errors in areas such as fronts. There are, however, two main disadvantages with isentropic coordinates: isentropic surfaces intersect with the ground in contrast to sigma coordinates and only statically stable solutions are allowed, since the vertical coordinate has to vary monotonically with height.

12.3 Structured and unstructured grids

The finite difference schemes mainly dealt with in this book are applied on structured grids such as those illustrated in Figure 12.7. The drawback of the Cartesian grids is that they do not change in space. Oceanic and atmospheric circulation models do not have Cartesian grids apart from some academic ones with pedagogical aims, cf. the present book. A typical grid used for practical purposes will at least have a latitude dependence of Δx, taking into account that the distances between longitudes decrease with increasing latitude. These grids are known as curvilinear, but are still orthogonal, *i.e.* preserve right angles between the two coordinates at every point of the grid. Figure 12.8 provides an example of this. The region bounded by two adjacent segments of one of the curvilinear coordinates and two adjacent segments of the other curvilinear coordinate will be transformable to a rectangle. An orthogonal curvilinear coordinate system permits the design of a grid system with the "north pole(s)" of the coordinate system shifted to a terrestrial location. Figure 12.8 shows a tri-polar grid, with the two "north poles" shifted to Siberia and the wastelands of northern Canada.

Unstructured grids, in contrast to the structured grids used with finite differences, do not require regular connectivity between the grid cells. The resolution can hence vary in space, with *e.g* higher resolution in coastal regions or narrow straits. Here, we will present a brief description of the general principles underlying the finite element and finite volume methods, which both employ unstructured grids.

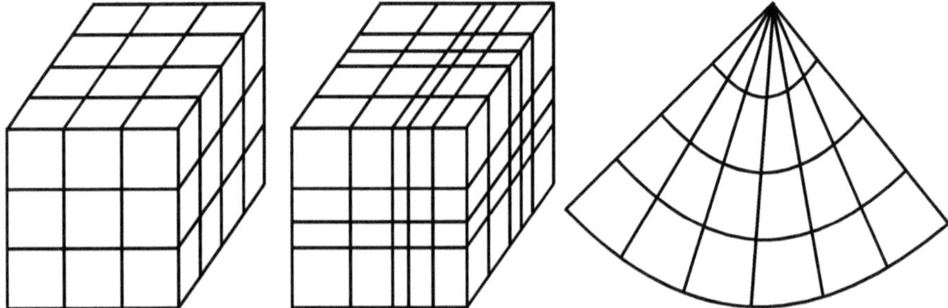

Figure 12.7. Three examples of structured grids. From left to right: 3D Cartesian, 3D rectilinear, 2D curvilinear. Note the "pole problem" of the curvilinear grid.

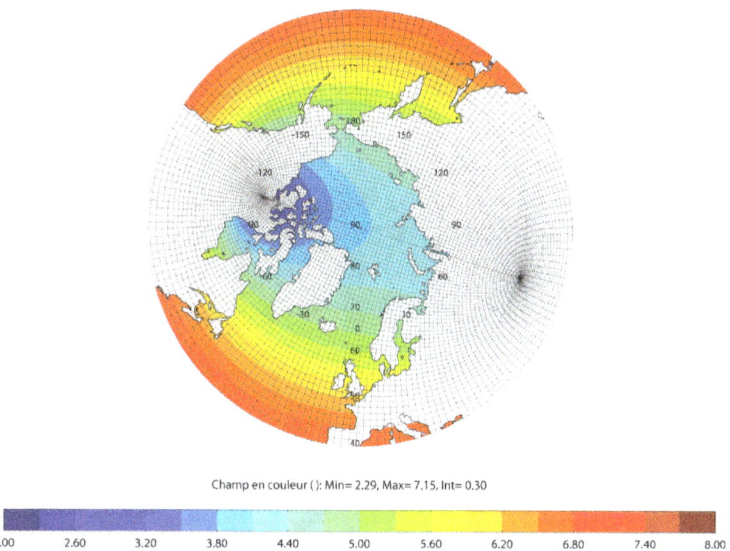

Figure 12.8. The orthogonal curvilinear ORCA12 ocean grid for the NEMO model, which is tripolar with two "north poles" in order to avoid the north pole being an ocean point. The colour scale indicates the grid size in km.

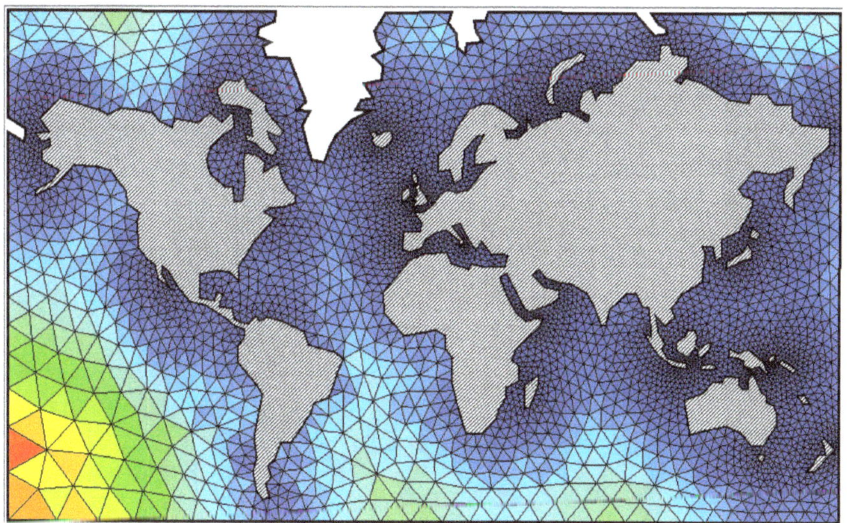

Figure 12.9. Finite elements for an ocean general circulation model

12.3.1 Finite element method

The main advantage of the finite element method is that one can locally increase the horizontal resolution, which is particularly important for coastal regions, cf. Figure 12.9. Disadvantages are, however, that the

CFL-criterion will be set by the smallest element, hereby imposing a severe limitation on the time step to be used in the integration. Furthermore data on an unstructured grid can be somewhat problematic to analyse.

The essence of the finite element method (FEM) is recognised by considering various ways of representing a function $u(x,t)$ on an interval $0 \leq x \leq L$. In the finite-difference method the function is defined only on a set of grid points; *i.e.* $u_j(t) \equiv u(x_j,t)$ is defined for a set of x_j, denoted nodes, but there is no explicit information about how the function behaves between these grid points.

In the finite element method, the function is defined in terms of a finite set of piecewise linear basis functions $e_j(x)$ as illustrated in Figure 12.10. The variable $u(x,t)$ is assumed to vary linearly between the nodes with a piecewise linear fit. The function $u(x,t)$ is then represented by the sum

$$u(x,t) = \sum_{j=0}^{N} u_j(t)e_j(x), \qquad (12.31)$$

where the grid is defined as $x_j = j\Delta x$ and $\Delta x = L/N$. The basis functions $e_j(x)$ are local, *i.e.* they are non-zero only on a small-sub-interval. Here the u_j are the coefficients of the basis functions and $u(x)$ is defined everywhere. The 1D linear advection equation employed in Chapter 4 to illustrate the use of finite differencing, will here be used to describe the idea underlying FEM:

$$\frac{\partial u}{\partial t} + c\frac{\partial u}{\partial x} = 0. \qquad (12.32)$$

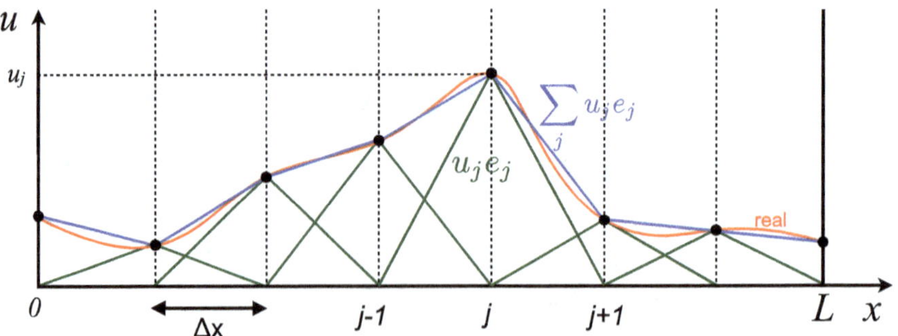

Figure 12.10. Illustration of how the variable u is built up with a combination of the piecewise linear basis functions e_j.

By inserting the piecewise linear representation of Equation (12.31) we obtain

$$\sum_j \frac{\partial u_j}{\partial t} e_j + c \sum_j u_j \frac{\partial e_j}{\partial x} = r, \tag{12.33}$$

where r is the residual. Point collocation, *viz.* setting $r = 0$, does not prove to be useful, but employing the Galerkin technique we may impose that

$$\int_0^L r e_i dx = 0 \tag{12.34}$$

for $i = 0, 1, 2, \ldots N$. Here r can be replaced using Equation (12.33), which yields

$$\sum_j \frac{\partial u_j}{\partial t} \int_0^L e_i e_j dx + c \sum_j u_j \int_0^L \frac{\partial e_j}{\partial x} e_i dx = 0. \tag{12.35}$$

The following results concerning the basis functions may be derived:

$$\int (e_{j+1} e_j) dx = \frac{\Delta x}{6} \ ; \ \int (e_{j \pm p} e_j) dx = 0 \ ; \ \int e_j^2 dx = \frac{2 \Delta x}{3} \ ; \tag{12.36a}$$

$$\int \frac{de_j}{dx} e_j dx = 0 \ ; \ \int \frac{de_{j \pm 1}}{dx} e_j dx = \pm \frac{1}{2} \ ; \ \int \frac{de_{j \pm p}}{dx} e_j dx = 0 \ , \tag{12.36b}$$

where p is any integer except $-1, 0, 1$. These relationships can now be used in order to reformulate Equation (12.35) as

$$\frac{1}{6} \left(\frac{du_{j+1}}{dt} + 4 \frac{du_j}{dt} + \frac{du_{j-1}}{dt} \right) + c \left(\frac{u_{j+1} - u_{j-1}}{2 \Delta x} \right) = 0. \tag{12.37}$$

The time derivative of u at the location x_j and time t^n is, for convenience, given by

$$G_j^n \equiv \frac{du_j}{dt}, \tag{12.38}$$

which, after substitution into Equation (12.37), yields

$$\frac{1}{6} \left(G_{j+1}^n + 4 G_j^n + G_{j-1}^n \right) = -c \left(\frac{u_{j+1} - u_{j-1}}{2 \Delta x} \right) . \tag{12.39}$$

The right-hand side of this equation is known and consequently it is possible to solve this system of simultaneous linear equations for all G_j^n

using the methods outlined in Chapter 10. Hereafter Equation (12.38) can be integrated in time using *e.g.* a leap-frog scheme:

$$u_j^{n+1} = u_j^{n-1} + 2\Delta t G_j^n.$$ (12.40)

The stability of this scheme can be analysed using the von Neumann method in the same manner as for the finite-difference schemes.

A comprehensive overview of oceanic finite element modelling can be found in Danilov et al. (2004) and a more fundamental one in Duben et al. (2012).

12.3.2 Finite volume method

Since the 1980s considerable scientific interest has increasingly been directed towards the finite volume method (FVM). Here, instead of dealing with variables at specific grid points (as in the case of finite differences), we consider their averages over given volumes. The philosophy behind FVM is that a volume integration of a PDE comprising a divergence term converts the latter to a surface integral of fluxes using the Gauss theorem. The finite volume method can be applied to both structured and unstructured grids. An advantage of FVM over FEM is that it conserves the variables better on a coarse grid.

As in the FEM case, we will use the 1D linear advection equation, here reformulated in "flux terms", to describe the general idea behind FVM:

$$\frac{\partial u}{\partial t} + \frac{\partial (cu)}{\partial x} = 0.$$ (12.41)

FVM uses averaged values of the variables over the given cells, rather than the values at specific grid points as in the finite-difference method. The cell $V_i^{n+\frac{1}{2}}$ describes the spatial region between $x_{i-\frac{1}{2}}$ and $x_{i+\frac{1}{2}}$ and the temporal region between t^n and t^{n+1} (cf. Figure 12.11). The spatially averaged value of u over this cell between $x_{i-\frac{1}{2}}$ and $x_{i+\frac{1}{2}}$ is

$$u_i^n = \frac{1}{\Delta x} \int_{x_{i-\frac{1}{2}}}^{x_{i+\frac{1}{2}}} u(x, t^n) \; dx.$$ (12.42)

The temporally averaged value of cu over this cell between t^n and t^{n+1} is

$$cu_i^{n+\frac{1}{2}} = \frac{1}{\Delta t} \int_{t^n}^{t^{n+1}} cu(x_i, t) \, dt.$$ (12.43)

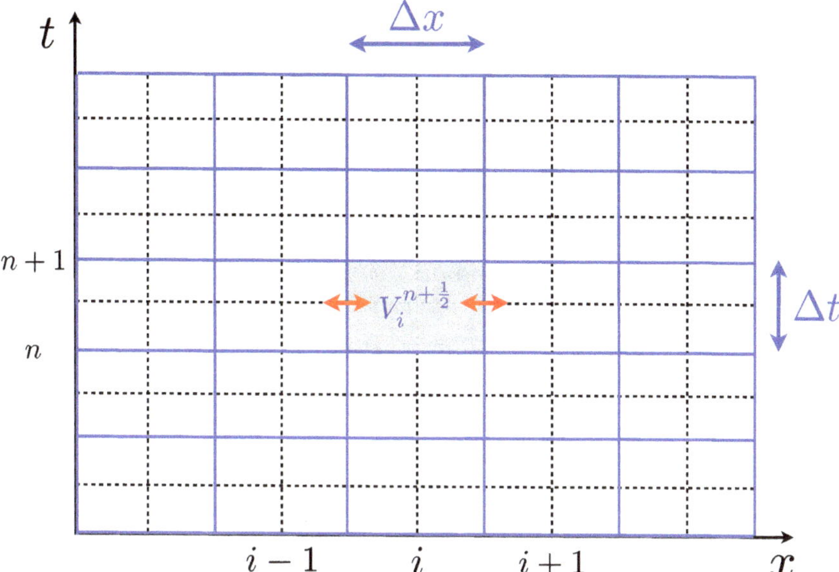

Figure 12.11. Spatiotemporal grid for a 1D finite volume model. The shaded cell $V_i^{n+\frac{1}{2}}$ is the spatial region between $x_{i-\frac{1}{2}}$ and $x_{i+\frac{1}{2}}$ and the temporal region between t^n and t^{n+1}. The "fluxes" in and out of this cell are indicated in red.

Integrating Equation (12.41) over a spatiotemporal cell and dividing by $\Delta x \Delta t$ we obtain

$$\frac{1}{\Delta x \Delta t} \int_{t^n}^{t^{n+1}} \int_{x_{i-\frac{1}{2}}}^{x_{i+\frac{1}{2}}} \frac{\partial u}{\partial t} \, dx \, dt + \frac{1}{\Delta x \Delta t} \int_{t^n}^{t^{n+1}} \int_{x_{i-\frac{1}{2}}}^{x_{i+\frac{1}{2}}} \frac{\partial (cu)}{\partial x} \, dx \, dt = 0.$$

$$(12.44)$$

After integration this becomes

$$\frac{1}{\Delta x \Delta t} \int_{x_{i-\frac{1}{2}}}^{x_{i+\frac{1}{2}}} \left(u^{n+1} - u^n \right) dx + \frac{1}{\Delta x \Delta t} \int_{t^n}^{t^{n+1}} \left(cu_{i+\frac{1}{2}} - cu_{i-\frac{1}{2}} \right) dt = 0.$$

$$(12.45)$$

This can be reformulated using Equations (12.42) and (12.43):

$$\frac{u_i^{n+1} - u_i^n}{\Delta t} + \frac{cu_{i+1/2}^{n+1/2} - cu_{i-1/2}^{n+1/2}}{\Delta x} = 0. \qquad (12.46)$$

Once the 1D edge "fluxes" have been calculated using either interpolation or extrapolation of the cell averages, this equation can be integrated forward in time:

$$u_i^{n+1} = u_i^n + \frac{c\Delta t}{\Delta x} \left(u_{i+1/2}^{n+1/2} - u_{i-1/2}^{n+1/2} \right). \tag{12.47}$$

Although this equation is superficially similar to what may be obtained using finite differences, it is important to underline that when physical space is multidimensional, the cells can be unstructured and the numerical integration clearly differs from that used when applying finite-difference methods. An overview of the use of the finite volume method in meteorology can be found in Machenhauer et al. (2009).

13. 3D Modelling

Ocean General Circulation Models are denoted OGCMs and Atmospheric General Circulation Models AGCMs. When coupled to each other they are often referred to as atmosphere-ocean coupled general circulation models (AOGCMs). Today's coupled climate models include not only circulation models but also other components such as land-surface, aerosols, chemistry, etc. and they are known as Earth System Models (ESMs). The core part is still the GCMs of the ocean and the atmosphere, which are increasing in complexity as modellers improve them. The numerical improvements can be such as higher-order numerical schemes, advanced model grids, etc. The improvement of the physics can be to replace the hydrostatic equations by the non-hydrostatic ones, to change the mixing parameterisations of the unresolved scales, etc. All these improvements are necessary in order to continue to undertake more realistic model integrations.

13.1 Approximations

A number of approximations and hypotheses are always necessary to make in order to simplify the full Navier-Stokes equations.

The spherical-Earth approximation

The geopotential surfaces are assumed to be spheres so that gravity, coinciding with the local vertical, is a constant. This approximation, instead of using the more accurate oblate spheroid, where the Earth "flattens" at the poles and "widens" at the Equator as a result of the centrifugal force. This approximation is made in all circulation models.

How to cite this book chapter:
Döös, K., Lundberg, P., and Campino, A. A. 2022. *Basic Numerical Methods in Meteorology and Oceanography*, pp. 135–147. Stockholm: Stockholm University Press. DOI: https://doi.org/10.16993/bbs.m. License: CC BY 4.0

The thin-layer approximation

The thickness of the atmosphere or the ocean is neglected compared to the radius of the Earth. This is also sometimes referred to as the shallow atmosphere/ocean approximation, which should not be confused with the approximations leading to the "shallow-water equations". The equations without the thin-layer approximation are known as the "deep equations" or the "non-hydrostatic deep equations".

The vertical Coriolis approximation

This approximation is sometimes confused with the thin-layer approximation or the hydrostatic approximation, but is in fact an approximation in itself. The $2\Omega w \cos\phi$ term in the zonal momentum equation and the $2\Omega u \cos\phi$ term in the vertical momentum equation are always omitted in hydrostatic models, but can be included in non-hydrostatic models, which will also be the case for the simple non-hydrostatic model presented later in this chapter.

The incompressibility approximation

The incompressibility approximation states that $D\rho/Dt = 0$. The mass conservation equation expressed by the continuity equation $D\rho/Dt + \rho\nabla \cdot \vec{V} = 0$ leads to $\nabla \cdot \vec{V} = 0$. The 3D divergence of the velocity is hence approximated to be zero and the volume is also conserved. An accurate expression for density computed by the equation of state may, however, be used when computing the pressure if the hydrostatic approximation is used. The fluid is therefore often referred to as pseudo-incompressible. This approximation is used in many OGCMs but never in AGCMs since air is highly compressible.

The Boussinesq approximation

Density variations are neglected except when they contribute to the buoyancy force, which explains the use of ρ_0 instead of ρ in the horizontal momentum equations (13.1a) and (13.1b). Since the density is still permitted to vary in the vertical momentum equation the approximation is therefore more accurately sometimes referred to as the "quasi-non-Boussinesq approximation". "Quasi" since the equation filters out the acoustic waves. The Boussinesq approximation is only good if the vertical variation of density is small relative to the mean density and that the horizontal and temporal variations are small relative to those in the vertical. This is consequently a good approximation for the ocean where $\rho \approx 1000 \, kg/m^3$.

It is important to note that this approximation and the incompressibility approximation both are consequences of the condition that the density variations be small compared to the mean density.

Non-Boussinesq approximations

The Boussinesq approximation can not be used for a realistic atmospheric model since the density decreases from $1.2 \, kg/m^3$ at sea level to $0 \, kg/m^3$ at the top of the atmosphere. It is instead possible to use other approximations such the anelastic or the pseudo-incompressible approximations in order to filter out acoustic waves.

The anelastic approximations are similar to the Boussinesq approximation but permit a vertical variation of the mean density so that $\rho = \rho_0(z) + \rho'(x, y, z, t)$, where $\rho_0(z)$ is the density satisfying the hydrostic balance and $\rho'(x, y, z, t)$ is a small perturbation around this balance. This eliminates sound waves by assuming that the flow has velocities much smaller than the speed of sound and permits a decreasing density with height. The anelastic approximations have, however, some important limitations since they deform the Rossby modes.

A more realistic approximation is the pseudo-incompressible approximation, which accounts for density fluctuations that arise from the the equation of state. Density fluctuations associated with perturbations in the pressure field are neglected. The pseudo-incompressible equation is the same as the anelastic continuity equation when the mean stratification is adiabatic. When the stability increases, however, the pseudo-incompressible approximation gives a more accurate result. The pseudo-incompressible approximation is, however, more exact when the stratification is stronger and the air more stable. The pseudo-incompressible approximation includes the effects of temperature changes on the density in the mass-conservation equation, which is not included in the anelastic approximation

The hydrostatic approximation

The vertical momentum equation is reduced to a balance between the vertical pressure gradient and the buoyancy force, which removes convective processes from the equations. Convection is instead parameterised with an increased vertical diffusion. This approximation is suitable for the large-scale circulation of the ocean and the atmosphere. But when the horizontal scale is shorter or on the same order as the depth scale this is not accurate anymore. We will, therefore, present a simple hydrostatic model as well as a non-hydrostatic model in this chapter. Note that use of pressure coordinates requires the hydrostatic approximation, cf. Section 12.2.2.

The turbulent-closure hypothesis

The urbulent-closure hypothesis is that the turbulent fluxes (representing the effect of small-scale processes on those of larger scales) are expressed in terms of large-scale features. In the present book this is most often expressed in terms of simple Laplacian diffusion and viscosity but can be parameterised in other ways. It is important to understand here that a model with higher resolution will not "need" as much sub-grid parameterisation as one of lower resolution since the former instead will resolve the scales better and "use" the hydrodynamic equations.

13.2 A simple hydrostatic model

In the present section, we will present a 3D circulation model formulated as simple as possible, using all the approximations above. A Cartesian grid will therefore be used, which is suitable for the ocean since depth is used as a vertical coordinate, this in contrast to atmospheric models employing pressure-dependent vertical coordinates. The numerical core of our 3D model will be close to that of a GCM. It should, however, be emphasised here that we do not recommend coding exactly these equations since their use would be very limited. The model is here discretised only for didactic reasons. Nearly all OGCMs are based on some type of curvilinear coordinates and depth-dependent layer thicknesses, which makes the discretised equations less transparent. We have therefore discretised the equations for a rectangular Cartesian C-grid as illustrated by Figures 13.1, 13.2 and 13.3.

Our simplified 3D-model is thus on a rectangular ocean domain with a flat bottom and furthermore makes use of the hydrostatic and Boussinesq approximations: The equations of motion that will be used are

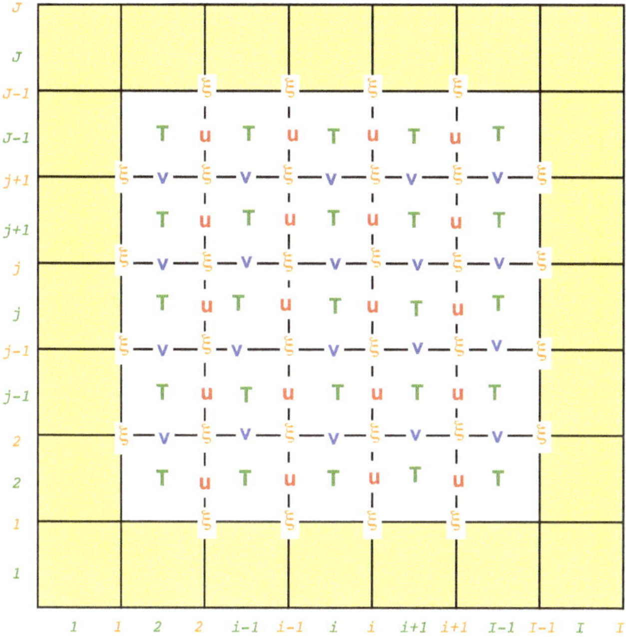

Figure 13.1. Horizontal view (longitude-latitude) of a possible rectangular model grid with land represented as yellow grid boxes. Only the non-zero variables are shown. I and J are the total number of grid boxes in the zonal and meridional direction, respectively.

$$\frac{\partial u}{\partial t} + u\frac{\partial u}{\partial x} + v\frac{\partial u}{\partial y} + w\frac{\partial u}{\partial z} - fv = -\frac{1}{\rho_0}\frac{\partial p}{\partial x} + A_H\nabla_H^2 u + A_V\frac{\partial^2 u}{\partial z^2} + F^x,$$

$$(13.1a)$$

$$\frac{\partial v}{\partial t} + u\frac{\partial v}{\partial x} + v\frac{\partial v}{\partial y} + w\frac{\partial v}{\partial z} + fu = -\frac{1}{\rho_0}\frac{\partial p}{\partial y} + A_H\nabla_H^2 v + A_V\frac{\partial^2 v}{\partial z^2} + F^y,$$

$$(13.1b)$$

$$0 = -\frac{\partial p}{\partial z} + \rho g,$$

$$(13.1c)$$

$$\frac{\partial u}{\partial x} + \frac{\partial v}{\partial y} + \frac{\partial w}{\partial z} = 0,$$

$$(13.1d)$$

$$\frac{\partial T}{\partial t} + u\frac{\partial T}{\partial x} + v\frac{\partial T}{\partial y} + w\frac{\partial T}{\partial z} = K_H\nabla_H^2 T + K_V\frac{\partial^2 T}{\partial z^2} + Q,$$

$$(13.1e)$$

$$\frac{\partial S}{\partial t} + u\frac{\partial S}{\partial x} + v\frac{\partial S}{\partial y} + w\frac{\partial S}{\partial z} = K_H\nabla_H^2 S + K_V\frac{\partial^2 S}{\partial z^2},$$

$$(13.1f)$$

$$\rho = \rho(S, T, p).$$

$$(13.1g)$$

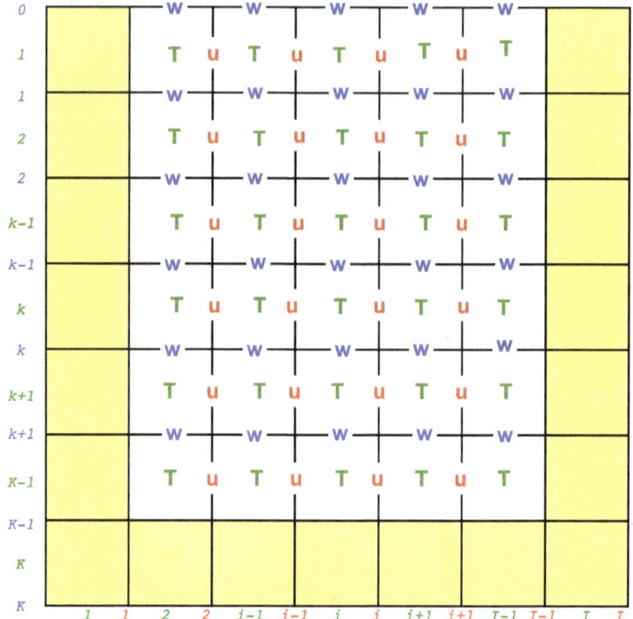

Figure 13.2. Zonal-vertical view of the model grid with land/sea floor represnted as yellow grid boxes. Only the non-zero variables are shown. K is the total number of vertical depth layers. Note that the vertical index k is increasing with depth and is hence in the of opposite direction of that of the z-coordinate.

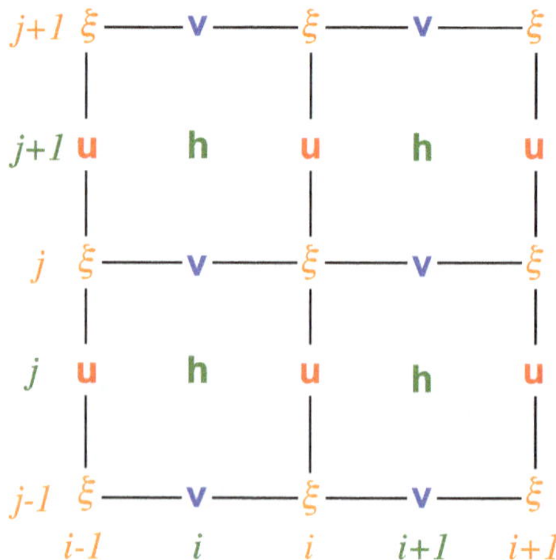

Figure 13.3. C-grid with points for the zonal velocity u, meridional velocity v, water or air column height h and vorticity ξ.

Here A and K are the viscosity and diffusion coefficients, respectively. The subscripts H and V pertain to horizontal and vertical processes, respectively. F^x and F^y represent the wind forcing, Q the thermal forcing at the sea surface. These seven equations for the ocean circulation correspond those established by Bjerknes (1904) for the atmosphere. The major differences are that the equation for humidity has been replaced by one for salinity and that air is compressible, whereas water has been taken to be incompressible. To prepare for discretisation it is convenient to rewrite the equations as

$$\frac{\partial u}{\partial t} = \xi v - w\frac{\partial u}{\partial z} - \frac{\partial E}{\partial x} + A_H \nabla_H^2 u + A_V \frac{\partial^2 u}{\partial z^2} + F^x, \tag{13.2a}$$

$$\frac{\partial v}{\partial t} = -\xi u - w\frac{\partial v}{\partial z} - \frac{\partial E}{\partial y} + A_H \nabla_H^2 v + A_V \frac{\partial^2 v}{\partial z^2} + F^y, \tag{13.2b}$$

$$\frac{\partial p}{\partial z} = \rho g, \tag{13.2c}$$

$$\frac{\partial w}{\partial z} = -\frac{\partial u}{\partial x} - \frac{\partial v}{\partial y}, \tag{13.2d}$$

$$\frac{\partial T}{\partial t} = -\frac{\partial(uT)}{\partial x} - \frac{\partial(vT)}{\partial y} - \frac{\partial(wT)}{\partial z} + K_H \nabla_H^2 T + K_V \frac{\partial^2 T}{\partial z^2} + Q, \tag{13.2e}$$

$$\frac{\partial S}{\partial t} = -\frac{\partial(uS)}{\partial x} - \frac{\partial(vS)}{\partial y} - \frac{\partial(wS)}{\partial z} + K_H \nabla_H^2 S + K_V \frac{\partial^2 S}{\partial z^2}, \tag{13.2f}$$

$$\rho = \rho(S, T, p), \tag{13.2g}$$

where the absolute vorticity has been defined as

$$\xi \equiv \frac{\partial v}{\partial x} - \frac{\partial u}{\partial y} + f \tag{13.3}$$

and the energy function as

$$E \equiv \frac{p}{\rho_0} + \frac{1}{2}\left(u^2 + v^2\right). \tag{13.4}$$

A possible discretisation of these equations with centred finite difference is

$$u_{i,j,k}^{n+1} = u_{i,j,k}^{n-1} + 2\Delta t$$

$$\times \left\{ \frac{1}{4}\left[\xi_{i,j,k}^n \left(v_{i,j,k}^n + v_{i+1,j,k}^n\right) + \xi_{i,j-1,k}^n \left(v_{i,j-1,k}^n + v_{i+1,j-1,k}^n\right)\right] \right.$$

$$- \left(w_{i,j,k}^n + w_{i+1,j,k}^n + w_{i,j,k-1}^n + w_{i+1,j,k-1}^n\right) \frac{u_{i,j,k-1}^n - u_{i,j,k+1}^n}{8\Delta z}$$

$$\left. - \frac{E_{i+1,j,k}^n - E_{i,j,k}^n}{\Delta x} \right\}, \tag{13.5a}$$

$$v_{i,j,k}^{n+1} = v_{i,j,k}^{n-1} + 2\Delta t$$
$$\times \left\{ -\frac{1}{4} \left[\xi_{i,j,k}^n \left(u_{i,j,k}^n + u_{i,j+1,k}^n \right) + \xi_{i-1,j,k}^n \left(u_{i-1,j,k}^n + u_{i-1,j+1,k}^n \right) \right] \right.$$
$$\left. - \left(w_{i,j,k}^n + w_{i,j+1,k}^n + w_{i,j,k-1}^n + w_{i,j+1,k-1}^n \right) \frac{v_{i,j,k-1}^n - v_{i,j,k+1}^n}{8\Delta z} \right.$$
$$\left. - \frac{E_{i,j+1,k}^n - E_{i,j,k}^n}{\Delta y} \right\},$$

$$(13.5b)$$

$$p_{i,j,k} = \sum_{k'=1}^{k-1} g\rho_{i,j,k'}\Delta z + g\rho_{i,j,k}\Delta z/2 + g\rho_{i,j,1}\eta_{i,j}, \qquad (13.5c)$$

$$w_{i,j,k-1} = w_{i,j,k} - \frac{U_{i,j} - U_{i-1,j}}{\Delta x} - \frac{V_{i,j} - V_{i,j-1}}{\Delta y}, \qquad (13.5d)$$

where the absolute vorticity is located between the corners of the T-boxes as illustrated by Figures 13.1 and 13.3:

$$\xi_{i,j,k} \equiv f + \frac{v_{i+1,j,k} - v_{i,j,k}}{\Delta x} - \frac{u_{i,j+1,k} - u_{i,j,k}}{\Delta y}. \qquad (13.6)$$

The fluxes U and V are defined at the same points as the velocity components u and v:

$$U_{i,j,k} \equiv u_{i,j,k}\frac{1}{2} \left(h_{i,j,k} + h_{i+1,j,k} \right), \qquad (13.7)$$

$$V_{i,j,k} \equiv v_{i,j,k}\frac{1}{2} \left(h_{i,j,k} + h_{i,j+1,k} \right). \qquad (13.8)$$

The grid-cell thickness is constant at all depths except at the surface, where the sea-surface elevation h is taken into account:

$$h_{i,j,k} = \Delta z + \eta_{i,j} \ for \ k = 1, \qquad (13.9)$$

$$h_{i,j,k} = \Delta z \ for \ k \neq 1. \qquad (13.10)$$

The gradient operator will act on the quantity E defined at the same locations as h:

$$E_{i,j,k} \equiv \frac{p_{i,j,k}}{\rho_0} + \frac{1}{2} \left[\frac{1}{2} \left(u_{i,j,k}^2 + u_{i-1,j,k}^2 \right) + \frac{1}{2} \left(v_{i,j,k}^2 + v_{i,j-1,k}^2 \right) \right]. \quad (13.11)$$

13.3 The tracer equation

The tracer equation describes the rate of change of a tracer such as *e.g.* potential temperature or salt in the ocean, water vapour in the

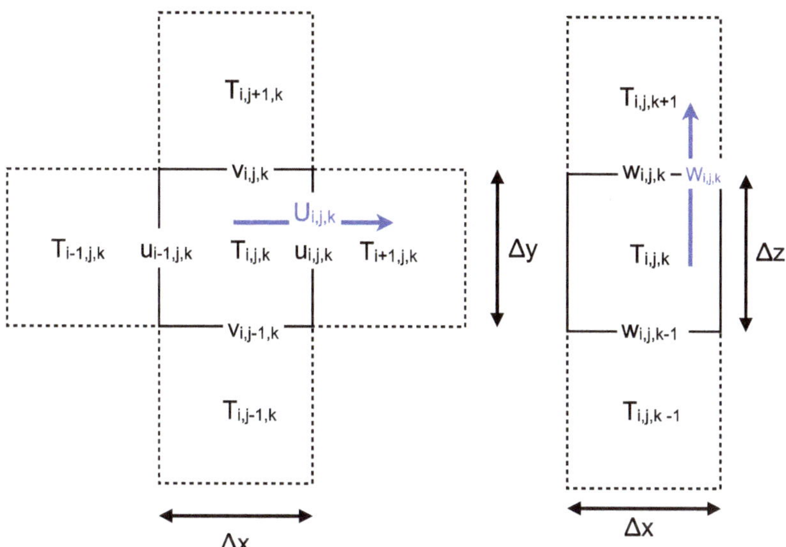

Figure 13.4. Horizontal (left) and vertical (right) views of the tracer equation applied on a C-grid. The blue arrows illustrate the tracer flux $U_{i,j,k} = u_{i,j,k}\frac{1}{2}\left(T_{i,j,k} + T_{i+1,j,k}\right)\Delta y\Delta z$ and $W_{i,j,k} \equiv w_{i,j,k}\frac{1}{2}\left(T_{i,j,k} + T_{i,j,k+1}\right)\Delta x\Delta y$.

atmosphere oɪ any tracer that is advected and diffused in the ocean or atmosphere. With a simple parameterisation of the diffusion the tracer equation can be expressed as

$$\frac{\partial T}{\partial t} + \vec{V}\cdot\nabla T = K_H\nabla_H^2 T + K_V\frac{\partial^2 T}{\partial z^2} + Q, \tag{13.12}$$

where K_H and K_V are the horizontal and vertical diffusion coefficients and Q a possible source term such as the heat flux between the atmosphere and the ocean. In the case of incompressibility, as postulated for our ocean model, the continuity equation is

$$\nabla\cdot\vec{V} = \frac{\partial u}{\partial x} + \frac{\partial v}{\partial y} + \frac{\partial w}{\partial z} = 0. \tag{13.13}$$

The tracer equation (13.12) can now be rewritten by incorporating the continuity equation (13.13):

$$\frac{\partial T}{\partial t} + \nabla\cdot\left(\vec{V}T\right) = K_H\nabla_H^2 T + K_V\frac{\partial^2 T}{\partial z^2} + Q. \tag{13.14}$$

13.3.1 Discretisation on a Cartesian grid

This tracer equation will now be discretised on the C-grid illustrated in Figure 13.4. Let us start with the term $\nabla \cdot \left(\vec{V} T \right)$ in Equation (13.14), which expresses the divergence of the tracer flux. The discretised version of this term is the sum of all the tracer transports in and out of a grid box divided by the volume of the grid box. The tracer flux across the grid wall where $u_{i,j,k}$ is located becomes

$$U^n_{i,j,k} \equiv u^n_{i,j,k} \frac{1}{2} \left(T^n_{i,j,k} + T^n_{i+1,j,k} \right) \Delta y \Delta z,$$

$$V^n_{i,j,k} \equiv v^n_{i,j,k} \frac{1}{2} \left(T^n_{i,j,k} + T^n_{i,j+1,k} \right) \Delta x \Delta z, \qquad (13.15)$$

$$W^n_{i,j,k} \equiv w^n_{i,j,k} \frac{1}{2} \left(T^n_{i,j,k} + T^n_{i,j,k+1} \right) \Delta x \Delta y,$$

which leads to

$$\nabla \cdot \left(\vec{V} T \right) \approx \frac{\left(U_{i,j,k} - U_{i-1,j,k} + V_{i,j,k} - V_{i,j-1,k} + W_{i,j,k-1} - W_{i,j,k} \right)}{\Delta x \Delta y \Delta z}.$$

$$(13.16)$$

With a centred leap-frog time scheme and the diffusion terms evaluated at time step $n-1$ as required to ensure stability, the discretised tracer equation becomes

$$
\begin{aligned}
T^{n+1}_{i,j,k} = T^{n-1}_{i,j,k} + \\
+ 2\Delta t \Bigg[&- \frac{U^n_{i,j,k} - U^n_{i-1,j,k} + V^n_{i,j,k} - V^n_{i,j-1,k} + W^n_{i,j,k-1} - W^n_{i,j,k}}{\Delta x \Delta y \Delta z} \\
&+ K_H \frac{T^{n-1}_{i-1,j,k} - 2T^{n-1}_{i,j,k} + T^{n-1}_{i+1,j,k}}{\left(\Delta x \right)^2} + \frac{T^{n-1}_{i,j-1,k} - 2T^{n-1}_{i,j,k} + T^{n-1}_{i,j+1,k}}{\left(\Delta y \right)^2} \\
&+ K_V \frac{T^{n-1}_{i,j,k-1} - 2T^{n-1}_{i,j,k} + T^{n-1}_{i,j,k+1}}{\left(\Delta z \right)^2} \Bigg) + Q^n_{i,j,k} \Bigg].
\end{aligned}
$$

$$(13.17)$$

13.3.2 Discretisation on an orthogonal curvilinear grid

A drawback of the discretised tracer equation above is that it requires Cartesian grids, which do not change in space. Oceanic and atmospheric circulation models do not have grids of this type apart from some academic ones used in courses on numerical methods.

It is therefore advantageous to apply finite differentiation directly to the diffusive tracer fluxes. Let U, V, W now instead be the sum of the advective and diffusive fluxes so that

$$U_{i,j,k}^n \equiv \left[u_{i,j,k}^n \frac{1}{2} \left(T_{i,j,k}^n + T_{i+1,j,k}^n \right) - K_H \frac{T_{i+1,j,k}^{n-1} - T_{i,j,k}^{n-1}}{\Delta x} \right] \Delta y_{i,j} \Delta z_k,$$

$$V_{i,j,k}^n \equiv \left[v_{i,j,k}^n \frac{1}{2} \left(T_{i,j,k}^n + T_{i,j+1,k}^n \right) - K_H \frac{T_{i,j+1,k}^{n-1} - T_{i,j,k}^{n-1}}{\Delta y} \right] \Delta x_{i,j} \Delta z_k,$$

$$W_{i,j,k}^n \equiv \left[w_{i,j,k+1}^n \frac{1}{2} \left(T_{i,j,k}^n + T_{i,j,k+1}^n \right) - K_{Vk} \frac{T_{i,j,k}^{n-1} - T_{i,j,k+1}^{n-1}}{\Delta z_k} \right] \Delta x_{i,j} \Delta y_{i,j}.$$

$$(13.18)$$

Note that we have written K_{Vk} with an added index k, which indicates that we can permit the vertical diffusion coefficient to vary with depth. It can be assumed to be either constant, or a function of the local Richardson number

$$R_i \equiv \frac{g}{\rho} \frac{\partial \rho}{\partial z} \bigg/ \left(\frac{\partial u}{\partial z} \right)^2 \tag{13.19}$$

or computed from a turbulent closure model using either the TKE or KPP formulation.

The discretised tracer equation now becomes

$$T_{i,j,k}^{n+1} = T_{i,j,k}^{n-1} -$$
$$2\Delta t \left(\frac{U_{i,j,k}^n - U_{i-1,j,k}^n + V_{i,j,k}^n - V_{i,j-1,k}^n + W_{i,j,k-1}^n - W_{i,j,k}^n}{\Delta x_{i,j} \Delta y_{i,j} \Delta z_k} - Q_{i,j,k} \right).$$

$$(13.20)$$

Note the additional horizontal indices on the grid lengths $\Delta x_{i,j}$ and $\Delta y_{i,j}$, this in order to conform to a curvilinear grid such as the one in Figure 12.8. The vertical grid thickness Δz_k will, however, only have a single vertical index since it solely varies in the vertical with the exception of the bottom grid boxes that might vary horizontally in order to fit an exact depth of the ocean.

13.4 Non-hydrostatic modelling

In a similar way as in the previous simple hydrostatic model we will here present an example of a simple non-hydrostatic model. This is based on the incompressible Boussinesq approximation of the equations of motion in z-coordinates. We still have the same seven variables (u, v, w, p, ρ, S, T), which are computed with the same set of seven equations but without the hydrostatic approximation and with the vertical Coriolis terms included.

$$\frac{\partial u}{\partial t} = -\frac{1}{\rho_0}\frac{\partial p}{\partial x} + G_x, \tag{13.21a}$$

$$\frac{\partial v}{\partial t} = -\frac{1}{\rho_0}\frac{\partial p}{\partial y} + G_y, \tag{13.21b}$$

$$\frac{\partial w}{\partial t} = -\frac{1}{\rho_0}\frac{\partial p}{\partial z} + G_z, \tag{13.21c}$$

$$\frac{\partial u}{\partial x} + \frac{\partial v}{\partial y} + \frac{\partial w}{\partial z} = 0, \tag{13.21d}$$

$$\frac{\partial T}{\partial t} = Q_T, \tag{13.21e}$$

$$\frac{\partial S}{\partial t} = Q_S, \tag{13.21f}$$

$$\rho = \rho(S, T, p), \tag{13.21g}$$

where

$$G_x = -u\frac{\partial u}{\partial x} - v\frac{\partial u}{\partial y} - w\frac{\partial u}{\partial z} + 2\Omega(v\sin\phi - w\cos\phi) + F^x, \tag{13.22a}$$

$$G_y = -u\frac{\partial v}{\partial x} - v\frac{\partial v}{\partial y} - w\frac{\partial v}{\partial z} - 2\Omega u\sin\phi + F^y, \tag{13.22b}$$

$$G_z = -u\frac{\partial w}{\partial x} - v\frac{\partial w}{\partial y} - w\frac{\partial w}{\partial z} + 2\Omega u\cos\phi + \frac{\rho g}{\rho_0} + F^z, \tag{13.22c}$$

$$Q_T = -u\frac{\partial T}{\partial x} - v\frac{\partial T}{\partial y} - w\frac{\partial T}{\partial z} + K_H\nabla_H^2 T + K_V\frac{\partial^2 T}{\partial z^2} + Q_A, \tag{13.22d}$$

$$Q_S = -u\frac{\partial S}{\partial x} - v\frac{\partial S}{\partial y} - w\frac{\partial S}{\partial z} + K_H\nabla_H^2 S + K_V\frac{\partial^2 S}{\partial z^2}. \tag{13.22e}$$

By taking

$$\frac{\partial}{\partial x}(13.21a) + \frac{\partial}{\partial y}(13.21b) + \frac{\partial}{\partial z}(13.21c) \tag{13.23}$$

and by using the continuity equation we obtain

$$\nabla^2 p = \rho_0 \nabla \cdot \vec{G}, \tag{13.24}$$

where $\vec{G} \equiv (G_x, G_y, G_z)$. This elliptic equation for pressure replaces the vertical integration of the hydrostatic equation (13.1b).

A difference when solving the governing equations with the non-hydrostatic terms included are that the vertical velocity is solved directly from the vertical momentum equation instead of integrating the continuity equation vertically. Furthermore the pressure is obtained by using an equation derived from the continuity equation together with the three momentum equations.

The fundamental difference between an oceanic non-hydrostatic model as the one presented here and an atmospheric one is that in the atmosphere one does not apply the Boussinesq approximation and that the fluid is compressible. The acoustic waves are therefore included since they are slow enough to be resolvable explicitly in the horizontal directions, whereas an implicit treatment in the vertical direction is sufficient to make the solution stable. Non-hydrostatic atmospheric models are essentially hyperbolic in pressure in contrast to the oceanic ones that are elliptic.

14. Spectral Methods

In some atmospheric general circulation models (AGCMs), the horizontal spatial representation of scalar dynamic and thermodynamic fields is based on truncated series of spherical harmonic functions. The nature of the underlying two-dimensional horizontal physical grid, also known as a transform grid, is tightly coupled to the parameters of the spherical harmonic expansion itself.

The numerical integration methods discussed so far are based on a discrete representation of the data on a grid of points encompassing the space over which a prediction of the variables is desired. Local time derivatives of the quantities to be predicted are determined by expressing the horizontal and vertical advection terms, sources etc. in finite-difference form. Finally, the time extrapolation is achieved by one of many possible algorithms, *e.g.* the leap-frog scheme. The finite-difference technique has a number of associated problems, such as truncation errors and linear as well as non-linear instabilities. Despite these difficulties, the finite-difference method has been the most practical method of generating forecasts numerically from the dynamical equations. As mentioned above there is another approach known as the spectral method which avoids some of the difficulties cited previously, in particular non-linear instability; however, this method is less versatile and the required computations are comparatively time-consuming. In a general sense, the mode of representation of data depends on their nature and the shape of the region over which the representation is desired. An alternative to depiction on a grid of discrete points is a representation in the form of a series of orthogonal functions. This requires the determination of the expansion coefficients of these functions,

How to cite this book chapter:
Döös, K., Lundberg, P., and Campino, A. A. 2022. *Basic Numerical Methods in Meteorology and Oceanography*, pp. 149–155. Stockholm: Stockholm University Press. DOI: https://doi.org/10.16993/bbs.n. License: CC BY 4.0

and is said to be a spectral representation, or a series expansion, in wave-number space. When such functions are used, the spatial derivatives can be evaluated analytically, eliminating the need for approximating them using finite differences.

Many AGCMs employ spectral techniques for their horizontal discretisations. This method has become the one most widely used for integrating the governing equations over hemispheric or global domains. The spectral method is, however, also used in some regional NWP models such as HARMONIE-AROME (Bengtsson et al., 2017).

One of the main virtues of the spectral technique is the combination with semi-implicit methods. Grid-point models have to use expensive methods to solve the elliptic balance equation required at every time step in a semi-implicit model. In spectral models this is instead inexpensive and accurate. Furthermore, semi-Lagrangian techniques avoid the calculation of nonlinear advection terms which is expensive in pure spectral models. Spectral transform models can compute advection terms less expensively. Finally the representation of positive definite quantities like moisture is an issue in spectral models.

Associated Legendre polynomials

The associated Legendre polynomials are the canonical solutions of the general Legendre equation

$$\frac{d}{dx}\left[(1-x^2)\frac{d}{dx}P_n^m(x)\right] = -n(n+1)P_n^m, \qquad (14.1)$$

where the indices n and m are referred to as the degree and order of the associated Legendre polynomial, respectively. The independent variable can be reparameterised in terms of angles, letting $x = cos\theta$, where $\theta = \pi/2 - \varphi$ is the colatitude. These polynomials to a low degree n and order m are presented in Table 14.1 and in Figure 14.1.

14.1 Spherical harmonics

As basis functions global atmospheric models use spherical harmonics (Tallqvist, 1905), which are the eigenfunctions of the Laplace equation on the sphere:

$$\nabla^2 Y_n^m = \frac{1}{a^2}\left[\frac{1}{cos^2\varphi}\frac{\partial^2 Y_n^m}{\partial\lambda^2} + \frac{1}{cos\varphi}\frac{\partial}{\partial\varphi}\left(cos\varphi\frac{\partial Y_n^m}{\partial\varphi}\right)\right] = -\frac{n(n+1)}{a^2}Y_n^m.$$

$$(14.2)$$

Table 14.1. The associated Legendre polynomials P_n^m in $x = cos\theta$, where $\theta = \pi/2 - \varphi$ is the colatitude.

	$m = 0$	$m = 1$	$m = 2$	$m = 3$
$n = 0$	1			
$n = 1$	$\cos\theta$	$-\sin\theta$		
$n = 2$	$\frac{1}{2}(3\cos^2\theta - 1)$	$-3\cos\theta\sin\theta$	$-3\sin^2\theta$	
$n = 3$	$\frac{1}{2}(5\cos^3\theta - 3\cos\theta)$	$-\frac{3}{2}(5\cos^2\theta - 1)\sin\theta$	$15\cos\theta\sin^2\theta$	$-15\sin^3\theta$
.				
.				

The spherical harmonics are products of Fourier series in longitude λ and associated Legendre polynomials in latitude φ:

$$Y_n^m(\lambda, \varphi) = P_n^m(\mu) e^{im\lambda}, \qquad (14.3)$$

where $\mu = sin\varphi$, m is the zonal wavenumber and n is the "total" wavenumber in spherical coordinates (as suggested by the Laplace equation).

In the usual application of the method, the basic prognostic variables are vorticity, divergence, temperature, a humidity variable, and the logarithm of surface pressure. Their horizontal representation is in terms of truncated series of spherical harmonic functions, whose variations are described by sines and cosines in the east-west direction and by associated Legendre polynomials from north to south. The horizontal variation of a variable U is thus given by

$$U(\lambda, \varphi, t) = \sum_{n=0}^{N} \sum_{m=-n}^{n} U_n^m(t) Y_n^m(\lambda, \varphi), \qquad (14.4)$$

where the spatial resolution is uniform over the sphere. This has a major advantage over finite differences based on a latitude-longitude grid, where the convergence of the meridians at the poles requires very small time steps. Although there are solutions for this "pole problem" for finite differences, the natural approach when solving the pole problem for global models is by using spherical harmonics.

It is becoming increasingly common to use what is known as the "triangular" truncation of this expansion (Figure 14.2). This is defined by $M = N = constant$, and yields a uniform resolution over the sphere. The symbol "TN" is the usual way of defining the resolution of

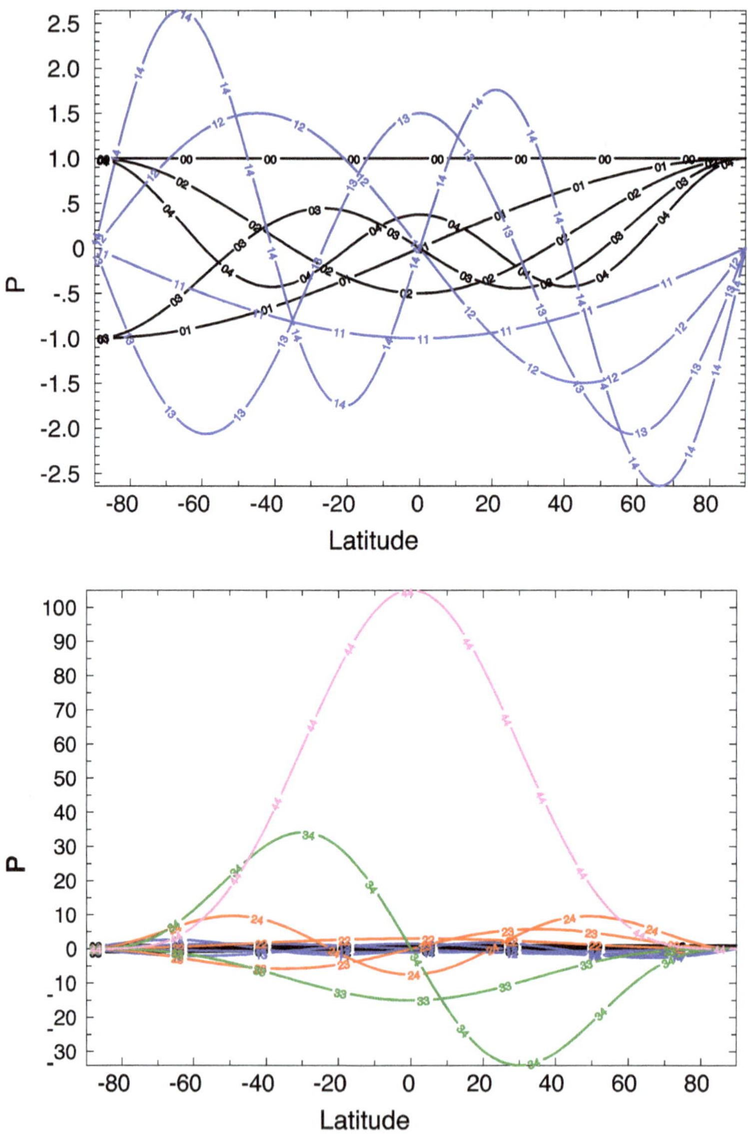

Figure 14.1. The first 15 associated Legendre polynomials P_n^m. The first number on the curves indicates the order of the polynomial and the second the degree. The upper panel shows these polynomials with $m = 1, 2$ for $n = 1, 2, 3, 4$. The lower panel shows the polynomials with $m = 1, 2, 3, 4$ for $n = 1, 2, 3, 4$. Each colour corresponds to a separate order m of the polynomial P_n^m.

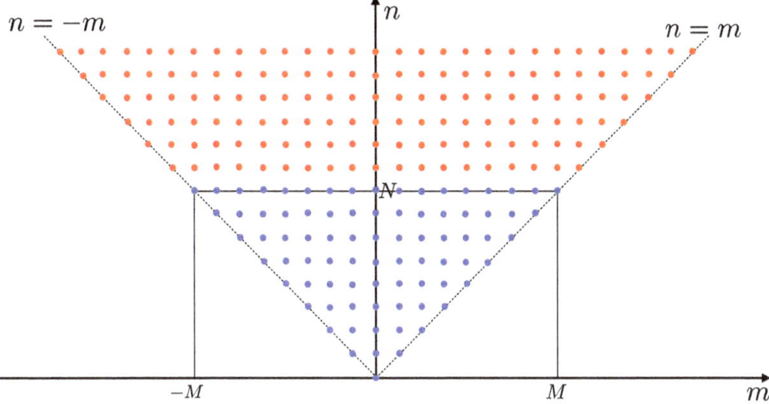

Figure 14.2. The triangular truncation in the (m, n) wavenumber space delimits a three-cornered region of spherical harmonic modes indicated by blue dots. Modes outside of this triangle are indicated with red dots.

such a truncation; N being the smallest "total" wave number retained in the expansion. Given that the Earth's radius is a, the smallest resolved half-wavelength in any particular direction is $\pi a/N$ (320 km for T63, 190 km for T106), although the corresponding lateral variation is of larger scale.

Derivatives of a spectrally represented variable U are given analytically:

$$\frac{\partial U}{\partial \lambda} = \sum_{n=0}^{N} \sum_{m=-n}^{n} imU_n^m Y_n^m$$

and

$$\frac{\partial U}{\partial \varphi} = \sum_{n=0}^{N} \sum_{m=-n}^{n} U_n^m \frac{\partial P_n^m}{\partial \varphi} e^{im\lambda}.$$

14.2 The spectral transform method

In a spectral model a variable, $\xi(\lambda, \varphi)$ is represented by a truncated series of spherical harmonic functions. This can be expressed as

$$\xi_n^m = \sum_{j=1}^{J} \xi^m(\varphi_j) P_n^m(\varphi_j), \tag{14.5}$$

where j is the latitudinal index, $P_n^m(\varphi_j)$ the associated Legendre polynomials and the expansion coefficients $\xi^m(\varphi_j)$ are obtained by a Fourier transform of $\xi(\lambda, \varphi)$.

The grid-point values are obtained by the inverse transform

$$\xi^m(\varphi) = \sum_{n=|m|}^{N(m)} \xi_n^m P_n^m(\varphi), \qquad (14.6)$$

followed by an inverse Fourier transform to obtain $\xi(\lambda, \varphi)$.

In a spectral model, the explicit time steps and evaluation of the horizontal gradients are undertaken in spectral space. The tendencies of the equations are, however, evaluated in grid-point space. One benefit when representing the variables in spectral space is that horizontal derivatives are represented in a continuous fashion, *i.e.* no finite differencing is needed to evaluate gradients. The methods used for the spherical harmonic transforms are, however, beyond the scope of this book. It is therefore suggested to use an already existing library to carry out the spectral transforms.

14.3 The shallow-water equations on a sphere

The momentum and mass continuity equations governing the motion of a rotating, homogenous, incompressible and hydrostatic fluid can be written in vector form as

$$\frac{d\vec{V}}{dt} = -f\mathbf{k} \times \vec{V} - \nabla\Phi + \nu\nabla^2\vec{V}, \qquad (14.7)$$

$$\frac{d\Phi}{dt} = -\Phi\nabla \cdot \vec{V}, \qquad (14.8)$$

where $\vec{V} = (u, v)$ is the horizontal velocity vector, Φ is the geopotential height, f is the Coriolis parameter and ν is the horizontal diffusion coefficient. Furthermore,

$$\frac{d}{dt} = \frac{\partial}{\partial t} + \vec{V} \cdot \nabla, \qquad (14.9)$$

and the ∇ operator is defined in spherical coordinates as

$$\nabla = \frac{1}{a\cos\varphi}\frac{\partial}{\partial\lambda} + \frac{1}{a}\frac{\partial}{\partial\varphi}, \qquad (14.10)$$

where a is the Earth's radius, λ is the longitude coordinate and φ is the latitude coordinate.

The equations above describe the shallow-water equations in the u, v, Φ system. In the model, we use another form of these equations. By introducing the relative vorticity ζ and horizontal divergence δ, the equations can be transformed into the the ζ, δ, Φ system. We do not derive these equations here since this procedure is relatively straightforward.

By introducing

$$\xi = \mathbf{k} \cdot (\nabla \times \vec{V}), \qquad (14.11)$$

and

$$\delta = \nabla \cdot \vec{V}, \qquad (14.12)$$

one can obtain the following set of equations (with $\mu = \sin \varphi$)

$$\frac{\partial \eta}{\partial t} = -\frac{1}{a(1 - \mu^2)} \frac{\partial}{\partial \lambda} U\eta + \frac{1}{a} \frac{\partial}{\partial \mu} V\eta, \qquad (14.13)$$

$$\frac{\partial \delta}{\partial t} = \frac{1}{a(1 - \mu^2)} \frac{\partial}{\partial \lambda} U\eta - \frac{1}{a} \frac{\partial}{\partial \mu} V\eta + \nabla^2 \left(\Phi + \frac{U^2 + V^2}{2(1 - \mu^2)} \right), \qquad (14.14)$$

$$\frac{\partial \Phi}{\partial t} = -\frac{1}{a(1 - \mu^2)} \frac{\partial}{\partial \lambda} U\Phi + \frac{1}{a} \frac{\partial}{\partial \mu} V\Phi, \qquad (14.15)$$

where $\eta = \xi + f$ is the absolute vorticity, which includes the Earth's rotation and $(U, V) = (u, v) \cos \varphi$.

15. Theoretical Exercises

The aim here is to apply the theory given in this book and to hereby provide a better understanding of basic numerical methods.

15.1 Exercises given in the main body of the text

These exercises are mainly found at the end of each pertinent chapter.

15.1.1 Finite differences

Determine the order of accuracy of the centred discretisations of the advection scheme

$$\frac{u_j^{n+1} - u_j^{n-1}}{2\Delta t} + c\frac{u_{j+1}^n - u_{j-1}^n}{2\Delta x} = 0. \tag{15.1}$$

Solution:
The Taylor series of $f(x)$ that is infinitely differentiable at x_0 is a power series:

$$f(x) = f(x_0) + (x - x_0)\left.\frac{\partial f}{\partial x}\right|_{x=x_0} + \frac{1}{2}(x - x_0)^2 \left.\frac{\partial^2 f}{\partial x^2}\right|_{x=x_0} + \mathcal{O}[(x - x_0)^3].$$

By defining $u(x_0, t_0) \equiv u_j^n$, the following Taylor series is obtained:

$$u^n_{j\overset{+}{-}1} \equiv u(x_0 \pm \Delta x, t_0) =$$

$$u(x_0, t_0) + (\cancel{x_0} \pm \Delta x - \cancel{x_0}) \left.\frac{\partial u}{\partial x}\right|_{x_0} + \frac{1}{2}(\cancel{x_0} \pm \Delta x - \cancel{x_0})^2 \left.\frac{\partial^2 u}{\partial x^2}\right|_{x_0}$$

How to cite this book chapter:
Döös, K., Lundberg, P., and Campino, A. A. 2022. *Basic Numerical Methods in Meteorology and Oceanography*, pp. 157–168. Stockholm: Stockholm University Press. DOI: https://doi.org/10.16993/bbs.o. License: CC BY 4.0

$$+\tfrac{1}{6}(x_0 \pm \Delta x - x_0)^3 \left.\frac{\partial^3 u}{\partial x^3}\right|_{x_0} + \mathcal{O}[(x_0 \pm \Delta x - x_0)^4]$$

$$= u(x_0, t_0) \pm \Delta x \left.\frac{\partial u}{\partial x}\right|_{x_0} + \tfrac{1}{2}(\Delta x)^2 \left.\frac{\partial^2 u}{\partial x^2}\right|_{x_0}$$

$$\pm \tfrac{1}{6}(\Delta x)^3 \left.\frac{\partial^3 u}{\partial x^3}\right|_{x_0} + \mathcal{O}[(\Delta x)^4]$$

and analogously

$$u_j^{n\pm 1} \equiv u(x_0, t_0 \pm \Delta t) = \tag{15.2}$$

$$u(x_0, t_0) + (t_0 \pm \Delta t - t_0) \left.\frac{\partial u}{\partial t}\right|_{t_0} + \tfrac{1}{2}(t_0 \pm \Delta t - t_0)^2 \left.\frac{\partial^2 u}{\partial t^2}\right|_{t_0}$$

$$+\tfrac{1}{6}(t_0 \pm \Delta t - t_0)^3 \left.\frac{\partial^3 u}{\partial t^3}\right|_{t_0} + \mathcal{O}[(t_0 \pm \Delta t - t_0)^4]$$

$$= u(x_0, t_0) \pm \Delta t \left.\frac{\partial u}{\partial t}\right|_{t_0} + \tfrac{1}{2}(\Delta t)^2 \left.\frac{\partial^2 u}{\partial t^2}\right|_{t_0}$$

$$\pm \tfrac{1}{6}(\Delta t)^3 \left.\frac{\partial^3 u}{\partial t^3}\right|_{t_0} + \mathcal{O}[(\Delta t)^4]. \tag{15.3}$$

The temporal derivative is

$$\frac{u_j^{n+1} - u_j^{n-1}}{2\Delta t} = \frac{1}{2\Delta t}\left(2\Delta t \left.\frac{\partial u}{\partial t}\right|_{t_0} + \frac{2}{6}(\Delta t)^3 \left.\frac{\partial^3 u}{\partial t^3}\right|_{t_0} + \mathcal{O}[(\Delta t)^4]\right)$$

$$= \left(\left.\frac{\partial u}{\partial t}\right|_{t_0} + \mathcal{O}[(\Delta t)^2]\right)$$

and the spatial derivative is

$$c\frac{u_{j+1}^n - u_{j-1}^n}{2\Delta x} = c\frac{1}{2\Delta x}\left(2\Delta x \left.\frac{\partial u}{\partial x}\right|_{x_0} + \frac{2}{6}(\Delta x)^3 \left.\frac{\partial^3 u}{\partial x^3}\right|_{x_0} + \mathcal{O}[(\Delta x)^4]\right)$$

$$= c\left(\left.\frac{\partial u}{\partial x}\right|_{x_0} + \mathcal{O}[(\Delta x)^2]\right).$$

Adding both of these expressions, the discretised advection equation is recovered:

$$\frac{u_j^{n+1} - u_j^{n-1}}{2\Delta t} + c\frac{u_{j+1}^n - u_{j-1}^n}{2\Delta x} = \left.\frac{\partial u}{\partial t}\right|_{t_0} + c\left.\frac{\partial u}{\partial x}\right|_{x_0} + \mathcal{O}[(\Delta x)^2, (\Delta t)^2].$$

From the analytical solution it is known that $\partial u/\partial t + c\,\partial u/\partial x = 0$. Therefore, the centred scheme in both time and space applied to the advection equation has an accuracy of $\mathcal{O}[(\Delta x)^2, (\Delta t)^2]$.

15.1.2 Stability Analysis

1. *Consider the leap-frog scheme for the advection equation:*

$$\frac{u_j^{n+1} - u_j^{n-1}}{2\Delta t} + c\frac{u_{j+1}^n - u_{j-1}^n}{2\Delta x} = 0. \tag{15.4}$$

   Use

$$u_j^n = \lambda^n e^{ikj\Delta x} u_0, \tag{15.5}$$

   and show that the amplification factor is

$$\lambda = -i\frac{c\Delta t}{\Delta x}\sin(k\Delta x) \pm \sqrt{1 - \left[\frac{c\Delta t}{\Delta x}\sin(k\Delta x)\right]^2}. \tag{15.6}$$

2. *Show that for $\mu > 1$ in the exercise above, one of the solutions of the differential equation will "blow up", at least for some wave lengths.*

Solution:
Introducing a wave solution into the discretised advection equation (15.4), the following expression obtained:

$$\left(\frac{\lambda - \lambda^{-1}}{2\Delta t} + c\frac{e^{ik\Delta x} - e^{-ik\Delta x}}{2\Delta x}\right) u_j^n = 0. \tag{15.7}$$

Rearranging the terms and looking for the non-trivial solution $u_j^n \neq 0$, it is found that,

$$\lambda^2 + 2ic\tfrac{\Delta t}{\Delta x}\sin(k\Delta x)\lambda - 1 = 0 \quad \rightarrow$$

$$\lambda = -i\mu\sin(k\Delta x) \pm \sqrt{1 - (\mu\sin(k\Delta x))^2},$$

where $\mu \equiv c\Delta t/\Delta x$ is the Courant number.

We will compute the norm of λ; there are two cases depending on the magnitude of μ:

$$\text{if } |\mu| \le 1,\ |\lambda|^2 = \lambda\lambda^* = \left(-ia \pm \sqrt{1-a^2}\right) \cdot \left(ia \pm \sqrt{1-a^2}\right) = 1,$$

if $|\mu| > 1$, λ is purely imaginary. Therefore,

$$|\lambda|^2 = 2\mu^2 \sin^2(k\Delta x) \tag{15.8}$$

$$- 1 \mp 2\mu \sin(k\Delta x)\sqrt{(\mu\sin(k\Delta x))^2 - 1}.$$

In the "worst" case $k\Delta x = \frac{\pi}{2}$, $|\lambda|^2 = 2\mu^2 - 1 \mp 2\mu\sqrt{\mu^2 - 1}$ and since $|\mu| > 1$, one of the roots is larger than one, and the solution will blow up.

3. *Discretise the advection equation with Euler-forward schemes in both time and space. Show that for $c > 0$ (backward scheme), the amplitude of the solutions will grow in time (these thus being unstable). But for $c < 0$ (forward scheme) the amplitude will decrease in time, i.e. the solution is stable.*

Solution:
The discretised advection equation using Euler-forward schemes in both time and space:

$$\frac{u_j^{n+1} - u_j^n}{\Delta t} + c\frac{u_{j+1}^n - u_j^n}{\Delta x} = 0. \tag{15.9}$$

Introducing a wave solution into the discretised equation yields

$$\left(\frac{\lambda - 1}{\Delta t} + c\frac{e^{ik\Delta x} - 1}{\Delta x}\right) u_j^n = 0. \tag{15.10}$$

Rearranging terms, an expression for λ is obtained:

$$\lambda = 1 - c\frac{\Delta t}{\Delta x}\left(e^{ik\Delta x} - 1\right). \tag{15.11}$$

In order to analyse the stability of the scheme, the squared norm of λ is calculated:

$$|\lambda|^2 = \left[1 - c\tfrac{\Delta t}{\Delta x}\left(e^{ik\Delta x} - 1\right)\right] \cdot \left[1 - c\tfrac{\Delta t}{\Delta x}\left(e^{-ik\Delta x} - 1\right)\right]$$

$$= 1 - \mu\left(e^{ik\Delta x} - 1\right) - \mu\left(e^{-ik\Delta x} - 1\right) + \mu^2\left(1 + 1 - e^{ik\Delta x} - e^{-ik\Delta x}\right)$$

$$= 1 - \mu\left[2\cos(k\Delta x) - 2\right] + \mu^2\left[2 - 2\cos(k\Delta x)\right]$$

$$= 1 - \left[2\cos(k\Delta x) - 2\right]\left(\mu + \mu^2\right)$$

$$= 1 + 4\sin^2\left(\tfrac{k\Delta x}{2}\right)\left(\mu + \mu^2\right) = |\lambda|^2.$$

The stability of the scheme is determined partly by the sign of the Courant number μ:

if $\mu > 0$, then $|\lambda|^2 > 1$ and the solution is unstable,

if $\mu < 0$, then we have two different cases. If $\mu \geq -1$ the solution is stable. However, if $\mu < -1$ the scheme becomes unstable. The stability criterion in this case becomes

$$-1 \leq c\frac{\Delta t}{\Delta x} \leq 0. \tag{15.12}$$

4. *Undertake a stability analysis of the following discretisation of the advection equation:*

$$\frac{u_j^{n+1} - \tfrac{1}{2}\left(u_{j+1}^n + u_{j-1}^n\right)}{\Delta t} + c\frac{u_{j+1}^n - u_{j-1}^n}{2\Delta x} = 0. \tag{15.13}$$

Solution:
Introducing the wave solution into the discretised advection equation yields

$$\left[\frac{\lambda - \tfrac{1}{2}\left(e^{ik\Delta x} + e^{-ik\Delta x}\right)}{\Delta t} + c\frac{e^{ik\Delta x} - e^{-ik\Delta x}}{2\Delta x}\right] u_j^n = 0. \tag{15.14}$$

Rearranging the terms and looking for the nontrivial solution $u_j^n \neq 0$, it is found that,

$$\lambda = \cos(k\Delta x) - ic\frac{\Delta t}{\Delta x}\sin(k\Delta x). \tag{15.15}$$

15.1.3 Accuracy of the numerical phase speed

Derive the numerical phase speed

$$C_D = \frac{1}{k\Delta t} \arcsin\left[\frac{c\Delta t}{\Delta x}\sin(k\Delta x)\right]. \tag{15.16}$$

Solution:
Start by considering the advection equation with centred schemes in both time and space:

$$\frac{u_j^{n+1} - u_j^{n-1}}{2\Delta t} + c\frac{u_{j+1}^n - u_{j-1}^n}{2\Delta x} = 0. \tag{15.17}$$

A wave solution of the form $u_j^n = u_0 e^{ik(j\Delta x - C_D n\Delta t)}$ is introduced into Equation (15.17).

The following relationships are to be employed:

$$u_j^{n+1} = u_0 e^{ik(j\Delta x - C_D(n+1)\Delta t)} = u_0 e^{-ikC_D\Delta t} e^{ik(j\Delta x - C_D n\Delta t)} = e^{-ikC_D\Delta t} u_j^n,$$

$$u_j^{n-1} = u_0 e^{ik(j\Delta x - C_D(n+1)\Delta t)} = u_0 e^{ikC_D\Delta t} e^{ik(j\Delta x - C_D n\Delta t)} = e^{ikC_D\Delta t} u_j^n,$$

$$u_{j+1}^n = u_0 e^{ik((j+1)\Delta x - C_D n\Delta t)} = u_0 e^{ik\Delta x} e^{ik(j\Delta x - C_D n\Delta t)} = e^{ik\Delta x} u_j^n,$$

$$u_{j-1}^n = u_0 e^{ik((j-1)\Delta x - C_D n\Delta t)} = u_0 e^{-ik\Delta x} e^{ik(j\Delta x - C_D n\Delta t)} = e^{-ik\Delta x} u_j^n.$$

We obtain the following expression:

$$\left(\frac{e^{-ikC_D\Delta t} - e^{ikC_D\Delta t}}{2\Delta t} + c\frac{e^{ik\Delta x} - e^{-ik\Delta x}}{2\Delta x}\right) u_j^n = 0. \tag{15.18}$$

The nontrivial solution is considered. Applying the trigonometrical identity $e^{ia} - e^{-ia} = 2i\sin(a)$ yields

$$\frac{-2i\sin(kC_D\Delta t)}{2\Delta t} + c\frac{2i\sin(k\Delta x)}{2\Delta x} = 0. \tag{15.19}$$

Rearranging terms,

$$\sin(kC_D\Delta t) = \frac{c\Delta t}{\Delta x}\sin(k\Delta x), \tag{15.20}$$

and applying the arcsin function to both sides of this equality, the final solution is obtained:

$$kC_D\Delta t = \arcsin\left[\frac{c\Delta t}{\Delta x}\sin(k\Delta x)\right]. \tag{15.21}$$

15.1.4 Diffusion and friction terms

1. *Undertake a stability analysis for the finite-difference version of the Rayleigh-friction equation with the right-hand side of Equation (8.3) taken at time step n.*

2. *Same as in step 1 but the right-hand side taken at time step n-1.*

3. *Same as in step 1 but the right-hand side taken at time step n+1.*

Solution:

Consider the following equation:

$$\frac{u_j^{n+1} - u_j^{n-1}}{2\Delta t} = -\gamma u_j^{n+a}, \tag{15.22}$$

where a can be either 0, -1 or 1. Insertion of a wave solution yields

$$\left(\frac{\lambda - \lambda^{-1}}{2\Delta t} + \gamma \lambda^a\right) u_j^n = 0. \tag{15.23}$$

After taking into account the nontrivial solution and rearranging terms, it is found that

$$\lambda^2 + 2\Delta t \gamma \lambda^{a+1} - 1 = 0. \tag{15.24}$$

Three cases are considered:

Case 1: $a = 0$

$$\lambda^2 + 2\Delta t \gamma \lambda - 1 = 0, \tag{15.25}$$

with the roots

$$\lambda_{1,2} = -\Delta t \gamma \pm \sqrt{(\Delta t \gamma)^2 + 1}. \tag{15.26}$$

One of these roots is characterised by $|\lambda| > 1$. The scheme is hence unconditionally unstable.

Case 2: $a = -1$

$$\lambda^2 + 2\Delta t \gamma - 1 = 0, \tag{15.27}$$

which has the roots

$$\lambda_{1,2} = \pm\sqrt{1 - 2\Delta t \gamma}. \tag{15.28}$$

The scheme is stable for $0 \leq \Delta t \gamma \leq 1$. However, for $1/2 \leq \Delta t \gamma \leq 1$ the roots become purely imaginary and the solution will be oscillatory, which is unphysical. The scheme is hence conditionally stable.

Case 3: $a = 1$

$$\lambda^2 + 2\Delta t \gamma \lambda^2 - 1 = 0, \tag{15.29}$$

where the roots satisfy

$$\lambda^2 = \frac{1}{1 + 2\Delta t \gamma}. \tag{15.30}$$

Both Δt and γ are positive, and hence $0 \leq \lambda^2 \leq 1$, and the scheme is thus unconditionally stable.

4. *Calculate the stability criterion for*

$$\frac{\partial u}{\partial t} = A \frac{\partial^2 u}{\partial x^2}, \tag{15.31}$$

using the following finite-difference scheme:

$$\frac{u_j^{n+1} - u_j^n}{\Delta t} = A \frac{u_{j+1}^n - 2u_j^n + u_{j-1}^n}{(\Delta x)^2}. \tag{15.32}$$

Estimate an upper limit for Δt when
$A = 10^6 \, m^2/s$, $\Delta x = 400 \, km$ (large-scale horizontal diffusion),
$A = 1 \, m^2/s$, $\Delta x = 10 \, m$ (vertical diffusion in a boundary layer).

Solution:
A wave solution is introduced into Equation (15.32), yielding

$$\left[\frac{\lambda - 1}{\Delta t} - A \frac{e^{ik\Delta x} - 2 + e^{-ik\Delta x}}{(\Delta x)^2} \right] u_j^n = 0. \tag{15.33}$$

The nontrivial solution is considered. Applying the trigonometrical identity $e^{ia} + e^{-ia} = 2\cos(a)$ yields

$$\lambda = 1 + A \frac{\Delta t}{(\Delta x)^2} [2\cos(k\Delta x) - 2] = 1 - 4A \frac{\Delta t}{(\Delta x)^2} \sin^2 \left(\frac{k\Delta x}{2} \right). \tag{15.34}$$

In the "worst" case $\sin^2(k\Delta x/2) = 1$. The stable non-oscillatory solution is given by the following condition:

$$0 \leq \Delta t \leq \frac{(\Delta x)^2}{4A}. \tag{15.35}$$

The results for the two cases are

$$A = 10^6\, m^2/s,\ \Delta x = 400\, km \rightarrow \Delta t^{max} = 11\, h.$$
$$A = 1\, m^2/s,\ \Delta x = 10\, m \rightarrow \Delta t^{max} = 25\, s.$$

5. *The diffusion equation can be integrated using the Crank-Nicholson scheme:*

$$\frac{T_j^{n+1} - T_j^n}{\Delta t} = \frac{A}{2}\left[\frac{T_{j+1}^n - 2T_j^n + T_{j-1}^n}{(\Delta x)^2} + \frac{T_{j+1}^{n+1} - 2T_j^{n+1} + T_{j-1}^{n+1}}{(\Delta x)^2}\right].$$

Examine the stability of this scheme!
Solution:
Introducing wave solutions into equation (15.36) yields

$$\left\{\frac{\lambda - 1}{\Delta t} - \frac{A}{2}\left[\frac{e^{ik\Delta x} - 2 + e^{-ik\Delta x}}{(\Delta x)^2} + \lambda\frac{e^{ik\Delta x} - 2 + e^{-ik\Delta x}}{(\Delta x)^2}\right]\right\}T_j^n = 0.$$

The nontrivial solution is considered by applying the trigonometrical identity $e^{ia} + e^{-ia} = 2\cos(a)$, and it is found that

$$\lambda\left[1 + 2\Delta t A \sin^2\left(\frac{k\Delta x}{2}\right)\right] = 1 - 2\Delta t A \sin^2\left(\frac{k\Delta x}{2}\right). \tag{15.36}$$

In simplified form:

$$\lambda = \frac{1 - 2\Delta t A \sin^2(k\Delta x/2)}{1 + 2\Delta t A \sin^2(k\Delta x/2)} \leq 1. \tag{15.37}$$

Even in the "worst" case $\sin^2(k\Delta x/2) = 1$, λ is always smaller than one. The scheme is thus unconditionally stable. However, the scheme is implicit.

15.2 Additional theoretical exercises

In contrast to the previous exercises, the students will need solve the following exercises by themselves.

15.2.1 Leap-frog scheme

Study the leap-frog scheme for the advection equation

1. Apply the leap-frog scheme (centered scheme with 2nd order accuracy in space and time).
2. Derive the stability criterion.
3. Discuss the computational mode and how it can be avoided.

15.2.2 Upwind scheme

Examine the upwind scheme for the advection equation.

1. Apply the upwind scheme (uncentered with 1st order accuracy in space and time).
2. Derive the stability condition.

15.2.3 Euler-forward scheme

Examine the Euler-forward scheme for the diffusion equation.

1. Derive the stability criterion for the Euler-forward scheme applied to the diffusion equation:

$$\frac{u_j^{n+1} - u_j^n}{\Delta t} = A \frac{u_{j+1}^n - 2u_j^n + u_{j-1}^n}{(\Delta x)^2}.$$

2. Discuss how the time step should be chosen. Why is it good to choose a smaller time step than that given by the stability criterion? Hint: Study how the amplification factor depends on the wavelengths, especially those of highest wavenumbers.
3. Do the oscillations that may appear, due to this scheme, represent a numerical mode?

15.2.4 Staggered vs. unstaggered grid

The following code lines are taken from a larger Fortran code that interpolates the temperature at a given position (described by a real pseudo-index xu). The original code works for an Arakawa C-grid. For simplicity we will consider the 1D case where this Arakawa grid is reduced to a 1D staggered grid. The xu is defined using u-points as a reference while the temperature is defined at an h-point. The square brackets enclosing dashed lines should be filled in with pertinent instructions!

```fortran
SUBROUTINE interp(xu,T)
!  computes temperature at the position
!  given by xu by interpolation
IMPLICIT NONE
INTEGER   :: ip, im
REAL   :: xu, ax, Tint
REAL, DIMENSION(:), INTENT(IN)   :: T
[ --- ]
ip = NINT(xu) + 1
im = NINT(xu)
ax = REAL(ip) - xu
Tint = T(im)*ax + T(ip)*(1-ax)
[ --- ]
RETURN
END SUBROUTINE
```

Consider the case $\texttt{xu} = 10.4$ and the temperature values: $\texttt{T(9)} = 5$, $\texttt{T(10)} = 10$, $\texttt{T(11)} = 20$.

1. Compute the interpolated temperature $\texttt{Tint}$ at $\texttt{xu}$ using the code above.
2. The result obtained for the interpolated temperature $\texttt{Tint}$ is wrong, why? "Fix" the code (explain the changes) and recalculate $\texttt{Tint}$.
3. Would the original interpolation scheme work correctly (for any $\texttt{xu}$ value) if an unstaggered grid was used instead?

15.2.5 Order of accuracy

Consider the continuity equation of the 1D shallow-water equations:

$$\frac{\partial h}{\partial t} = -H\frac{\partial u}{\partial x}. \tag{15.38}$$

1. Discretise this equation with centred finite differences on an unstaggered grid.
2. Derive the order of accuracy of the two finite-difference schemes.
3. Repeat the previous tasks using instead a staggered grid.

15.2.6 Nonrotating 2D shallow-water equations

Consider the nonrotating 2D shallow-water equations.

1. Discretise these equation on an Arakawa B-grid and apply the leap-frog scheme in time.

2. Compute the stability criterion for these discretisations. For simplicity consider the case $\Delta x = \Delta y$.
3. Repeat steps 1 and 2 for the Arakawa C-grid.

15.2.7 Laplace equation

Consider the Laplace equation $\nabla^2 \Phi = 0$

1. Discretise this equation in space.
2. Set $\Delta x = \Delta y$ and write the simplest iterative scheme (Jacobi iteration).
3. Set up a 4×4 grid, with $\Phi = 1$ at the boundaries and iterate 3 times with a starting state of $\Phi = 0$.
4. Repeat the exercise using the faster Gauss-Seidel scheme.
5. Repeat the exercise using the even faster SOR scheme.

15.2.8 Semi-implicit scheme

Consider the nonrotating and y-independent shallow-water equations:

$$\frac{\partial u}{\partial t} = -g\frac{\partial h}{\partial x}, \tag{15.39}$$

$$\frac{\partial v}{\partial t} = 0, \tag{15.40}$$

$$\frac{\partial h}{\partial t} = -H\frac{\partial u}{\partial x}, \tag{15.41}$$

1. Discretise these equations on a C-grid using a semi-implicit scheme centred at time level n. Terms containing spatial partial derivatives must be evaluated as the average at time levels $(n-1)$ and $(n+1)$.
2. Undertake a stability analysis of these discretised equations.

16. GFD Computer Exercises

The aim of this chapter is to apply our previously gained skills in numerical methods to some fundamental problems in geophysical fluid dynamics (GFD).

16.1 Advection and diffusion equations

The leap-frog, the Euler-forward and the upwind schemes will be studied and applied to the advection and diffusion equations. The advection equation is

$$\frac{\partial u}{\partial t} + c\frac{\partial u}{\partial x} = 0, \tag{16.1}$$

and the diffusion equation is

$$\frac{\partial u}{\partial t} - A\frac{\partial^2 u}{\partial x^2} = 0. \tag{16.2}$$

The solution interval will in all cases be taken as $0 \le x \le 1$, with the periodic boundary condition $u(0,t) = u(1,t)$.

Relative error

The relative error E^n as a function of time level n is defined as

$$E^n = \left(\frac{\sum_{i=1}^{I} \left(u_i^n - \hat{u}_i^n\right)^2}{\sum_{i=1}^{I} \left(\hat{u}_i^n\right)^2} \right)^{1/2}, \tag{16.3}$$

where $\hat{u}_i^n$ is the discretised form of the analytical solution and I is the number of grid points in the spatial domain.

How to cite this book chapter:
Döös, K., Lundberg, P., and Campino, A. A. 2022. *Basic Numerical Methods in Meteorology and Oceanography*, pp. 169–185. Stockholm: Stockholm University Press. DOI: https://doi.org/10.16993/bbs.p. License: CC BY 4.0

16.1.1 Advection equation

Examine the the advection equation using a simple numerical model. Write a program that can solve the advection equation (set $c = 1$) with two different schemes: the leap-frog scheme and the upwind scheme.

a) Cosine wave:

    i) Model set-up: Run the program with the resolution $\Delta x = 0.02$ and the Courant numbers 0.9, 1.0 and 1.1. Use a cosine wave as initial condition:

$$u(x, t = 0) = \cos(2\pi x).$$

    ii) Solve the problem using the leap-frog scheme as well as the upwind schemes separately. Initialise the leap-frog scheme with a single Euler-forward step.

    iii) Plot the results obtained with both schemes and the analytic solution in the same figure.

    iv) Analyse the phase error and the amplitude error for both schemes. Do the results agree with the theory? (*Note:* C_D can be calculated analytically as the system has a single known value for k).

    v) Show how the relative error develops in time (consider running the code for as long times as $t \approx 10^4$).

b) Cosine pulse:

    i) Model set-up: Run the program with the resolution $\Delta x = 0.01$ and the Courant number 0.9. Use a cosine pulse as initial condition:

$$u(x, t = 0) = \begin{cases} \frac{1}{2} + \frac{1}{2}\cos\left(10\pi(x - 0.5)\right) & \text{for } 0.4 \leq x \leq 0.6, \\ 0 & \text{elsewhere.} \end{cases}$$

    ii) Solve the problem using a) the leap-frog scheme and b) the upwind scheme. Initialise the leap-frog scheme with a single Euler-forward step.

    iii) Try to identify the computational mode.

16.1.2 Diffusion equation

Examine the diffusion equation using a simple numerical model. Write a program that can solve the diffusion equation using the Euler-forward scheme. Set $\Delta x = 0.05$, $A_H = 1$, and try with three different time steps: on determined by the critical value of the von Neumann number $\nu \equiv A\Delta t/(\Delta x)^2$ permitted by the stability criterion, one slightly larger and one half of the maximum possible value.

Run the program with two different initial conditions:

i) Rectangular pulse:

$$u_i^1 = \begin{cases} 1 & \text{for } 0 \le x \le 0.5, \\ 0 & \text{for } 0.5 < x < 1. \end{cases}$$

ii) Spike: $u = 1$ at the single grid point $x = 0.5$ and $u = 0$ at all other grid points.

16.2 1D shallow-water model

Here we will examine the difference between implementing a staggered and an unstaggered grid applied to the 1D shallow-water equations as well as how to incorporate open boundary conditions in the model.

The shallow-water equations in the one-dimensional case are

$$\frac{\partial u}{\partial t} = -g\frac{\partial h}{\partial x}, \tag{16.4}$$

$$\frac{\partial h}{\partial t} = -H\frac{\partial u}{\partial x}. \tag{16.5}$$

These equations can be combined into a single expression for u and h:

$$\frac{\partial^2 u}{\partial t^2} - gH\frac{\partial^2 u}{\partial x^2} = \frac{\partial^2 h}{\partial t^2} - gH\frac{\partial^2 h}{\partial x^2} = 0. \tag{16.6}$$

The general d'Alembert solution to the wave equation is given by two functions travelling in opposite directions:

$$u(x,t) = u_1(x - ct) + u_2(x + ct), \tag{16.7}$$
$$h(x,t) = h_1(x - ct) + h_2(x + ct), \tag{16.8}$$

where $c = \sqrt{gH}$ is the propagation speed of the waves. The solution interval to be analysed here is $0 \le x \le 1$, with the periodic boundary condition $u(0,t) = u(1,t)$.

Examine one-dimensional gravity waves, which can be described by the shallow-water equations. Write a program that solves the 1D shallow-water equations given above.

a) Unstaggered grid:

   i) Model set-up: Discretise using the leap-frog scheme on an unstaggered grid. Set $H = g = 1$ (thus $c = \sqrt{gH} = 1$) to simplify the system. Run the program with $\Delta x = 0.025$ and the Courant number 0.9.

   ii) Initialise the leap-frog scheme with a single Euler-forward step, and use the following initial conditions:

$$h(x, t = 0) = \begin{cases} \frac{1}{2} + \frac{1}{2}\cos\left(10\pi(x - 0.5)\right) & \text{for } 0.4 \leq x \leq 0.6, \\ 0 & \text{elsewhere.} \end{cases}$$

$$u(x, t = 0) = 0.$$

   iii) Do the results agree with the analytical solution of the problem?

b) Staggered grid:

   i) Model set-up: Rewrite the program using a staggered grid, with h and u defined at different points.

   ii) Run the program and try to find the stability limit setting the time step. Then choose a time step 10 % smaller than that in the stability limit, and employ the same initial condition as in case a).

   iii) Run three different simulations: one with the resolution given in case a), a second one with half that resolution (finer grid) and a third one with double the case-a) resolution (coarser grid).

   iv) Select one of the variables (either h or u) and compare the accuracy of the solution with the previous solution obtained on an A-grid. Discuss the difference between the three results and the result from case a). Note that Δx is the distance between two h-points (and also two u-points).

c) Open boundary conditions:

   In order to gain an understanding of how to deal with open boundary conditions requiring a "sponge" (relaxation) zone we will start by examining the 1D case. Consider a relaxation zone defined in the region $0 \leq x \leq L_r$. The relaxation function γ has the following values at the boundaries of the sponge zone:

$$\gamma(L_r) = 0, \tag{16.9}$$

$$\gamma(0) = 1. \tag{16.10}$$

Here we shall consider two types of relaxation functions:

Linear function:

$$\gamma(x) = 1 - \frac{x}{L_r}. \tag{16.11}$$

Trigonometric function:

$$\gamma(x) = \frac{1}{2}\left[1 + \cos\left(\frac{\pi x}{L_r}\right)\right] \tag{16.12}$$

i) Model set-up: Consider the 1D shallow-water model on a domain $-L \leq x \leq L$ on a staggered grid. Set $c = L = 1$ and $\Delta x = 0.025$. Consider as initial condition a centrally located cosine-shaped "bump":

$$(h(x,0), u(x,0)) = (h_0, u_0)\begin{cases} \frac{1}{2} + \frac{1}{2}\cos\left(\frac{10\pi x}{L}\right) & \text{for } -\frac{L}{10} \leq x \leq \frac{L}{10} \\ 0 & \text{elsewhere} \end{cases}$$

Here $h_0 = u_0 = 1$. Impose a solid boundary at $x = L$ and an open boundary at $x = -L$ by adding a relaxation zone adjacent to the western boundary. The relaxation can be parameterised in the following way:

$$u_d(x) = \left[1 - \gamma(x)\frac{\Delta t}{\tau}\right] u(x) \quad \text{for} \quad -L_r \leq x \leq -L \tag{16.13}$$

where τ is a relaxation time-scale (s^{-1}) and u_d is the damping value. Construct the function so that $\gamma = 0$ in the interior of the system, and $\gamma \to 1$ as you approach the boundary. The function may be either trigonometric or linear. Here it is sufficient to set $\tau = \Delta t$ for simplicity.

ii) Analyse the effects of the length of the relaxation zone L_r, choose values within the interval $0 \leq L_r \leq L$. Select a proper magnitude to study the effect of the open boundary.

iii) Discuss the results and the differences you observe when you impose the trigonometric and the linear relaxation function.

16.3 2D shallow-water model

Here the aim is to solve the linearised 2D shallow-water equations using some standard numerical methods and to study atmospheric and oceanic processes. These equations, including rotation and physical parameterisations, are

$$\frac{\partial u}{\partial t} = fv - g\frac{\partial h}{\partial x} + F^x, \tag{16.14}$$

$$\frac{\partial v}{\partial t} = -fu - g\frac{\partial h}{\partial y} + F^y, \tag{16.15}$$

$$\frac{\partial h}{\partial t} = -H\left(\frac{\partial u}{\partial x} + \frac{\partial v}{\partial y}\right). \tag{16.16}$$

Here the physical parameterisations F^x and F^y may for instance be of the horizontal viscosity and/or the drag forcing.

Below you will find the general structure of a Fortran-coded shallow-water model. The dashed lines should be filled in with relevant Fortran instructions!

```fortran
PROGRAM structure_of_code
!------------------------------------------------------------!
! General structure of a shallow-water model in
! Fortran with staggered grid, rotation, diffusion
! & relaxation schemes.
! It is good and common practice to always write a
! few lines about what the code does and how.
!------------------------------------------------------------
IMPLICIT NONE
!------------------------------------------------------------
! Think about:
!    Explain each parameter in words and units.
!    Use common notation.
!    Do not write too much in one line.
! Physical constants:
REAL*4, PARAMETER :: f = ? ! Coriolis parameter [s-1]
REAL*4, PARAMETER :: g = ?  ! Gravity  [m s-2]
! Model parameters:
REAL*4, PARAMETER :: H  = ?  ! Mean depth [m]
REAL*4, PARAMETER :: mu = ? ! Diffusion coeff. [m2 s
   -1]
! Grid
INTEGER*4, PARAMETER  ::  NX = ?  ! Number of i-
   points
INTEGER*4, PARAMETER  ::  NY = ?  ! Number of j-
   points
INTEGER*4, PARAMETER  ::  NT = ? ! Number of time
   steps
REAL*4, PARAMETER :: dx = ?  ! Zonal grid spacing [m]
```

```fortran
REAL*4, PARAMETER :: dy = ?   ! Merid. grid spacing [m
   ]
REAL*4, PARAMETER :: dt = ?   ! Time step [s]
! Save data to file
CHARACTER*200 :: outFile = ? ! Name of output file
! Work variables
INTEGER*4 ::  ic, ip, im, jc, jp, jm, nc, nm, np
REAL*4 ::  du, dv, dh
! Data matrices
REAL*4, DIMENSION(NX,NY,NT) :: u, v, h
!----------------------------------------------------
! Initial condition
!----------------------------------------------------
! Main loop
! You might want to indent the code inside the loop
   to
! make it easier to see where the loop starts and
   ends.
DO nc=------
   np = nc+1
   nm = nc-1
   DO jc=-------
      jp = jc+1
      jm = jc-1
      DO ic=-------
         ip = ic+1
         im = ic-1
         ! Reset
         du = 0.
         dv = 0.
         dh = 0.
         ! Coriolis
         du = du + -------------------
         dv = dv + -------------------
         ! Sea surface height gradient
         du = du + -------------------
         dv = dv + -------------------
         ! Continuity
         dh = dh + -------------------
         ! Diffusion
         du = du + -------------------
         dv = dv + -------------------
         ! Rolaxation
         du = du + -------------------
         dv = dv + -------------------
         ! New time step
```

```fortran
        u(ic,jc,nt) = ------------------------------
        v(ic,jc,nt) = ------------------------------
        h(ic,jc,nt) = ------------------------------
        ! Asselin filtering
        u(ic,jc,nc) = ------------------------------
        v(ic,jc,nc) = ------------------------------
        h(ic,jc,nc) = ------------------------------
        ! Update time index
      ENDDO
      ! Don't forget to take care of boundary
    conditions,
      ! zonally periodic, no-slip, or other.
    ENDDO
ENDDO
!---------------------------------------------------------
! Save the data to file
!---------------------------------------------------------
END PROGRAM
```

The following tasks should be undertaken:

i) Model set-up: Here we develop a 2D shallow-water model that includes rotation on an Arakawa C-grid. Set F^x and F^y to zero. For simplicity prescribe your domain as $-1 \leq x, y \leq 1$ where $\Delta x = \Delta y = 0.025, g = 1$, and $H = 1$. Choose a proper Courant number and introduce an Asselin filter. Integrate the first time step using an Euler-forward scheme and the subsequent time steps with a leap-frog scheme and keep in mind that periodic boundaries will affect the computation of the Coriolis terms.

ii) Consider as initial conditions:

$$u(x,y,0) = 0, \tag{16.17}$$

$$v(x,y,0) = 0, \tag{16.18}$$

$$h(x,y,0) = h_0 e^{-(x/L_W)^2 - (y/L_W)^2}, \tag{16.19}$$

where $L = 1$, $L_W = L/7$ and $h_0 = 1$. Impose solid boundaries everywhere.

iii) Run the model and analyse the results for the non-rotating case $(f = 0)$.

iv) Calculate the potential and kinetic energies of the system. Study the time evolution of both their magnitudes and check to what extent the total energy is preserved. Does the Asselin filtering affect this result?

v) Repeat the same tasks for the case of constant rotation, $f = 10^{-4}s^{-1}$. Interpret the results physically.

Note that for each of the subsequent tasks given in this chapter you will need to programme new parts in the model code to deal with different physical processes such as waves and circulation in the ocean and atmosphere.

16.4 Geostrophic adjustment

The purpose here is to understand what controls the evolution of a disturbance initially at rest on a rotating plane, a process commonly known as geostrophic adjustment. To study this phenomenon, you will use the 2D shallow-water model developed in the previous section.

$$\frac{\partial u}{\partial t} - f_0 v = -g\frac{\partial h}{\partial x}, \tag{16.20}$$

$$\frac{\partial v}{\partial t} + f_0 u = -g\frac{\partial h}{\partial y}, \tag{16.21}$$

$$\frac{\partial h}{\partial t} + D_0 \left(\frac{\partial u}{\partial x} + \frac{\partial v}{\partial y} \right) = 0. \tag{16.22}$$

Realistic values of g, D_0 and f should be used in order to gain understanding of real atmospheric and oceanographic processes. The following should be programmed in the code:

i) Use parameters that are appropriate to the mid-latitude North Atlantic when running the experiments, if nothing else is given, viz. $L = 5\cdot 10^6$ m, $D_0 = 4000$ m, $g = 9.81$ ms^{-2}, $f_0 = 10^{-4}$ s^{-1}.

ii) Prescribe the number of grid cells N_X and N_Y in the x- and y-directions respectively.

iii) Let Δx, Δy be determined by L and N_X, N_Y .

iv) Let Δt be determined by a Courant number, phase speed and $\max(\Delta x, \Delta y)$.

v) Let N_T be determined by $T_{\max}$, which is the simulation period.

vi) Construct the main time loop so that the model does not store the fields every time step.

vii) Analyse the effect of the boundary condition by prescribing different widths of the sponge zone.

16.4.1 Geostrophic adjustment for a step-function disturbance

Here you will study the geostrophic adjustment for an initial disturbance consisting of a discontinuity in h that is symmetric in y.

Model set-up:

i) Implement a disturbance in h, which is described by a step function in the zonal direction. A good practice is to programme in Fortran the initial condition as a `CASE` or `LOGICAL` to be able to turn it on or off for other exercises.

ii) The code already describes a Gaussian disturbance of the sea-surface height as an initial condition, do not remove this, instead make this as one of the options for the initial condition.

iii) Use periodic boundaries by attaching the northern boundary to the southern.

iv) Use open boundaries in east and west. Make sure that the sponge zone is wide enough.

Exercises:

i) Derive the final steady state of the sea-surface height and the velocities starting from the linearised shallow-water equations.

ii) Run the model until the system only varies very little in time (steady state).

iii) Verify your model results with the results you obtained analytically.

iv) Tweak the parameters. Which parameters are most important?

v) Discuss the results obtained by tweaking the parameters.

16.4.2 Geostrophic adjustment for a Gaussian disturbance

Here you will study the geostrophic adjustment of an initial disturbance described by a Gaussian perturbation of the sea-surface height. You are supposed to study the various energetics of the system.

Model set-up:

i) Use a Gaussian disturbance instead of the step function employed in the previous subsection. The disturbance should be smaller than the domain, but make sure it is larger than the resolution.

ii) Programme the potential, kinetic and total energies of the system. Use the Fortran command **LOGICAL** to be able to turn the computation on or off. Remember that you have to save these fields to be able to study them.

iii) Use open boundaries (sponge) everywhere.

Exercise 1:

i) Write down the analytical expression for the total kinetic and potential energies.

ii) Run the model with two different depths; one shallow $D_0 \sim 500\ m$ and the other deep $D_0 \sim 10000\ m$.

iii) Study how the energies of the system (kinetic, potential and total) vary in time for the two cases with different depths.

iv) Give a physical explanation of the results.

Exercise 2:

The system can be considered to be in steady state when the energy of the system changes very little with time.

i) Run the model long enough so that the system reaches to a steady state for the depths used in Exercise 1.

ii) Run the model again using different depths and sizes of the disturbance.

iii) How do the results change? Try to explain the changes and compare with theory.

16.5 Kelvin wave

The aim of this assignment is to gain an understanding of the influence solid boundaries exert on geophysical flows. Although applicable to some situations in the atmosphere, the influence of lateral boundaries is more important in the world oceans. In this assignment you will use a simple shallow-water model with solid boundaries, and look at how an initial disturbance evolves, and how a system initially at rest evolves as you add forcing.

16.5.1 Gaussian disturbance

Model set-up:

i) Programme an infinite coast by setting an open boundary on the north side of the domain and a solid boundary on the south side.

ii) Impose periodic boundary conditions in the x-direction.

iii) Programme a new initial condition that describes a reasonably-sized Gaussian disturbance of the h-field centered in the middle of the coast, so that the partitioned disturbance has its maximum value at the wall.

iv) Introduce a reduced gravity in the system.

Experiment:

i) Run the model for two different cases; the North Atlantic ($L = 1 \cdot 10^7\ m$, $H = 1000\ m$) and the Baltic Sea ($L = 1 \cdot 10^6\ m$, $H = 30\ m$).

ii) Do any particular kind of waves show up?

iii) How can you identify these waves? (phase speed, shape, ...).

iv) Connect your results to the theory.

v) Run the model again for both cases (North Atlantic and the Baltic Sea) using different settings, $e.g.$ changing the resolution.

vi) Look at a cross-section of the wave, and estimate its phase speed.

vii) Compare your results with the theory.

16.5.2 Equatorial β-plane

In this part of the assignment you are supposed to study waves that appear on an equatorial β-plane. Model set-up:

i) Change the initial condition to a Gaussian disturbance centred in the middle of the domain.

ii) Programme a new Coriolis parameter so that it corresponds to an equatorial β-plane ($f = \beta y$). Try to take advantage of the Fortran command **LOGICAL** here so that you easily can change between a β-plane or an f-plane. Place the equator ($y = 0$) centrally in the domain. Remember to introduce reduced gravity in the code!

Experiment:

   i) Run the model using values corresponding to the Pacific Ocean
($L = 2 \cdot 10^7$ m , $H = 1000$ m)

   ii) Why do you need a reduced gravity for this simulation?

   iii) Identify the waves in the model.

   iv) Do the waves correspond to theory?

   v) Can these waves be observed in the real ocean?

16.6 Oceanic Rossby waves

The aim of this assignment is to study and understand the kind of waves
that are generated when $D(y)$ or $f(y)$ are "sloping", *viz.* $D = D_0 + \alpha y$ or
$f = f_0 + \beta y$. For this assignment you should use a model with periodic
and open (sponge) boundary conditions. The initial disturbance should
be in geostrophic balance to avoid gravity waves.

Model set-up:

   i) Use an initial disturbance in geostrophic balance. (Hint: you
have already programmed the initial condition for h in a
previous exercise. But you need to programme u and v so that
they are in geostrophic balance.)

   ii) Use periodic boundaries in x by attaching the eastern
boundary to the western.

   iii) Use open boundaries (sponge) on the southern and northern
boundaries.

   iv) Programme a sloping α-plane $D = D_0 + \alpha y$ using `LOGICAL` (or
`CASE`) so that you can choose to have it on or off in the
simulation.

   v) Programme a new Coriolis parameter $f = f_0 + \beta y$. Use the
Fortran command `LOGICAL` (or `CASE`) so that you can choose
to turn it on or off.

16.6.1 Rossby waves on a β-plane

   i) Consider a rectangular basin with $L = 7 \cdot 10^6$ m and
$D_0 = 4000$ m in the mid latitudes (*e.g.* the North Pacific).
Run the model for at least 30 days.

   ii) Start by deriving the phase speed and group velocity for
Rossby waves in this linear system. (The derivation should not
be included in the report, only the final solution).

iii) Run the model with a β-plane and a constant topography D_0.
iv) Describe and explain the evolution of the system.
v) Connect the results to theory.
vi) What kind of waves develop?
vii) Do they have any distinguishing properties?

16.6.2 Phase and group velocities

i) Rerun the model but with different wave numbers. (Hint: To change the wavenumber, change the width of the disturbance).
ii) Compare the obtained phase speed and group velocity to the theoretical values.
iii) Repeat this exercise but keep f constant and vary the topography using an α-plane. Discuss the differences between the results.

16.6.3 $\beta - \alpha$ compensation

i) Run the model using a varying topography and f-field $(D = D_0 + \alpha y$ and $f = f_0 + \beta y)$.
ii) Calculate the value of α that cancels the β effect. What happens to the initial disturbance under these circumstances?

16.7 Atmospheric Rossby waves

The aim of this assignment is to study and understand the kind of waves that are generated when $f(y)$ is "sloping", *viz.* $f = f_0 + \beta y$. The effect of including a zonal mean flow will also be discussed. For this assignment you should use a model with periodic and open (sponge) boundary conditions. The initial disturbance should be in geostrophic balance to avoid gravity waves.

Model set-up:

i) Use an initial disturbance in geostrophic balance. (Hint: you have already programmed the initial condition for h in a previous exercise. But you need to programme u and v so that they are in geostrophic balance.)
ii) Use periodic boundaries in x by attaching the eastern boundary to the western.

iii) Use open boundaries (sponge) on the southern and northern boundaries.

iv) Introduce a zonal mean flow U_0 in the system using **LOGICAL** (or **CASE**) so that you can choose to have it on or off in the simulation.

v) Programme a new Coriolis parameter describing $f = f_0 + \beta y$. Use **LOGICAL** (or **CASE**) so that you can choose to turn it on or off.

16.7.1 $\beta-$plane

i) Consider a rectangular basin with $L = 7 \cdot 10^6 \ m$ and $H = 4000 \ m$ in the mid latitudes. Run the model for at least 30 days.

ii) Start by deriving the phase speed and group velocity for Rossby waves in this linear system. (The derivation should not be included in the report, only the final solution).

iii) Run the model with a β-plane and without a mean flow $(U_0 = 0 \ m/s)$.

iv) Describe and explain the evolution of the system.

v) Connect the results to theory.

vi) What kind of waves develop?

vii) Do they have any distinguishing properties?

16.7.2 Phase and group velocities

i) Run the model as above, but with different wave numbers. (Hint: To change the wavenumber, change the width of the disturbance).

ii) Compare the obtained phase speed and group velocity to the theoretical values.

16.7.3 The effect of the zonal mean

i) Increase the zonal length of the domain to $L_x = 28 \cdot 10^6 \ m$ while using the same Δx as in the previous experiments.

ii) Run the model using a β plane and four different zonal mean flows (choose zonal flows in the range $0 < U_0 \leq 15 \ m/s$). Calculate the group velocity for each case.

iii) Using the results from ii) and try to find the value of the zonal mean flow at which the Rossby wave becomes stationary. Is the initial condition preserved? If not, explain why.

16.8 Gyre Circulations

The aim of this assignment is to study the Sverdrup, Stommel and Munk theories for barotropic large-scale ocean circulation. To study this phenomenon, you will use the 2D shallow-water model including friction:

$$\frac{\partial u}{\partial t} = fv - g\frac{\partial h}{\partial x} + A_H \nabla^2 u + \Gamma u, \tag{16.23}$$

$$\frac{\partial v}{\partial t} = -fu - g\frac{\partial h}{\partial y} + A_H \nabla^2 v + \Gamma v, \tag{16.24}$$

$$\frac{\partial h}{\partial t} = -H\left(\frac{\partial u}{\partial x} + \frac{\partial v}{\partial y}\right), \tag{16.25}$$

where A_H is the horizontal viscosity coefficient and Γ the Rayleigh-friction coefficient.

Model set-up:

  i) Let the system initially be at rest.
 ii) Use solid boundaries.
iii) Programme a wind forcing that is constant in x and sinusoidal in y.
 iv) Use a β-plane

Do not forget to compute the barotropic stream function in order to compare your results with the analytical solution.

16.8.1 Sverdrup solution

  i) Run the model without any friction and use parameters that are appropriate for the North Atlantic.
 ii) What effect does $\beta = \partial f/\partial y$ have on the system?
iii) What effect does the wind forcing have?
 iv) Do the results resemble theory? Why/Why not?
  v) Do the results improve if the size of the domain is increased? (*e.g.* rerun the experiment for the North Pacific)
 vi) What are the limitations of the Sverdrup theory?

16.8.2 Stommel solution

  i) Now run the model using the same ocean basin (either the North Atlantic or the North Pacific) with the Rayleigh linear friction applied **SOLELY** to v.

ii) Give a physical explanation of this kind of friction.

iii) Describe and explain the modelled circulation using the Stommel theory.

iv) Rerun the model using a different value of the Rayleigh-friction coefficient Γ.

v) Describe how the results have changed.

16.8.3 Munk solution

i) Now run the model using the same ocean basin (either the North Atlantic or the North Pacific) with Laplacian friction.

ii) Give a physical explanation of this kind of friction.

iii) Describe and explain the modelled circulation using the Munk theory.

iv) Rerun the model using different values of the horizontal viscosity coefficient A_H.

v) Describe how the results have changed.

vi) Compare the Stommel and Munk solutions!

Bibliography

Adcroft, A. and Campin, J.-M.: Rescaled height coordinates for accurate representation of free-surface flows in ocean circulation models, Ocean Modelling, doi:10.1016/j.ocemod.2003.09.003, 2004.

Asselin, R.: Frequency filter for time integrations, Monthly Weather Review, 100, 487–490, 1972.

Batteen, M. L. and Han, Y.-J.: On the computational noise of finite-difference schemes used in ocean models, Tellus, 33, 387–396, 1981.

Bengtsson, L., Andrae, U., Aspelien, T., Batrak, Y., Calvo, J., de Rooy, W., Gleeson, E., Hansen-Sass, B., Homleid, M., Hortal, M., et al.: The HARMONIE–AROME model configuration in the ALADIN–HIRLAM NWP system, Monthly Weather Review, 145, 1919–1935, doi:10.1175/: MWR-D-16-0417.1, 2017.

Bjerknes, V.: Das Problem der Wettervorhersage, betrachtet vom Standpunkte der Mechanik und der Physik, vol. 21, Meteorologische Zeitschrift, 1904.

Bryan, K. and Cox, M. D.: A numerical investigation of the oceanic general circulation, Tellus A, 19, 1967.

Charney, J. G., Fjørtoft, R., and von Neumann, J.: Numerical integration of the barotropic vorticity equation, Tellus, 2, 237–254, 1950.

Courant, R., Friedrichs, K., and Lewy, H.: Über die partiellen Differenzengleichungen der mathematischen Physik, Mathematische Annalen, 100, 32–74, 1928.

Courant, R., Isaacson, E., and Rees, M.: On the solution of nonlinear hyperbolic differential equations by finite differences, Communications on Pure and Applied Mathematics, 5, 243–255, doi:10.1002/cpa.3160050303, 1952.

Crank, J. and Nicolson, P.: A practical method for numerical evaluation of solutions of partial differential equations of the heat-conduction type, in: Mathematical Proceedings of the Cambridge Philosophical Society, vol. 43, pp. 50–67, Cambridge University Press, 1947.

Danilov, S., Kivman, G., and Schröter, J.: A finite-element ocean model: Principles and evaluation, Ocean Modelling, 6, 125–150, doi:10.1016/ S1463-5003(02)00063-X, 2004.

Döös, B. R. and Eaton, M. A.: Upper-Air Analysis over Ocean Areas, Tellus, 9, 184–194, 1957.

Duben, P. D., Korn, P., and Aizinger, V.: A discontinuous/continuous low order finite element shallow water model on the sphere, Journal of Computational Physics, 231, 2396–2413, doi:https://doi.org/10.1016/j.jcp.2011.11.018, 2012.

Durran, D.: Numerical Methods for Fluid Dynamics - With Applications to Geophysics, Texts in Applied Mathematics, Springer New York, 2010.

Fischer, G.: Ein numerisches Verfahren zur Errechnung von Windstau und Gezeiten in Randmeeren, Tellus, 11, 60–76, 1959.

Flather, R. A.: A Storm Surge Prediction Model for the Northern Bay of Bengal with Application to the Cyclone Disaster in April 1991, Journal of Physical Oceanography, 24, 172–190, doi:10.1175/1520-0485(1994)024 <0172:ASSPMF>2.0.CO;2, 1994.

Hackbusch, W.: Multi-Grid Methods and Applications, vol. 4, doi:10.1007/978-3-662-02427-0, 1985.

Hansen, W.: Theorie zur Errechnung des Wasserstandes und der Strömungen in Randmeeren nebst Anwendungen, Tellus, 8, 287–300, 1956.

Higdon, R. L.: Numerical Absorbing Boundary Conditions for the Wave Equation, Mathematics of Computation, 49, 65–90, URL http://www.jstor.org/stable/2008250, 1987.

Holton, J. and Hakim, G.: An Introduction to Dynamic Meteorology, Academic Press, Elsevier Science, 2013.

Hovmöller, E.: The trough-and-ridge diagram, Tellus, 1, 62–66, 1949.

Kalnay, E.: Atmospheric Modeling, Data Assimilation and Predictability, Cambridge University Press, 2003.

Killworth, P. D., Webb, D. J., Stainforth, D., and Paterson, S. M.: The development of a free-surface Bryan–Cox–Semtner ocean model, Journal of Physical Oceanography, 21, 1333–1348, 1991.

Kwizak, M. and Robert, A. J.: A semi-implicit scheme for grid point atmospheric models of the primitive equations, Monthly Weather Review, 99, 32–36, 1971.

Lynch, P.: The Emergence of Numerical Weather Prediction: Richardson's Dream, Cambridge University Press, 2006.

Machenhauer, B., Kaas, E., and Lauritzen, P. H.: Finite-Volume Methods in Meteorology, in: Special Volume: Computational Methods for the Atmosphere and the Oceans, edited by Temam, R. M. and Tribbia, J. J., vol. 14 of *Handbook of Numerical Analysis*, Elsevier, 2009.

Manabe, S. and Bryan, K.: Climate calculations with a combined ocean-atmosphere model, Journal of the Atmospheric Sciences, 26, 786–789, 1969.

Mesinger, F. and Arakawa, A.: Numerical methods used in atmospheric models, GARP technical report 17, WMO/ICSU. Geneva, Switzerland, 1, 1976.

Nycander, J. and Döös, K.: Open boundary conditions for barotropic waves, Journal of Geophysical Research: Oceans, 108, 2003.

Orlanski, I.: A simple boundary condition for unbounded hyperbolic flows, Journal of Computational Physics, 21, 251–269, 1976.

Pacanowski, R. C. and Gnanadesikan, A.: Transient response in a z-level ocean model that resolves topography with partial cells, Monthly Weather Review, 126, 3248–3270, 1998.

Persson, A.: Early operational Numerical Weather Prediction outside the USA: an historical Introduction. Part I: Internationalism and engineering NWP in Sweden, 1952–69, Meteorological Applications, 12, 135–159, 2005a.

Persson, A.: Early operational Numerical Weather Prediction outside the USA: an historical introduction: Part II: Twenty countries around the world, Meteorological Applications, 12, 269–289, 2005b.

Persson, A.: Early operational Numerical Weather Prediction outside the USA: an historical introduction Part III: Endurance and mathematics–British NWP, 1948–1965, Meteorological Applications, 12, 381–413, 2005c.

Phillips, N. A.: A coordinate system having some special advantages for numerical forecasting, Journal of Meteorology, 14, 184–185, 1957.

Richardson, L. F.: The Approximate Arithmetical Solution by Finite Differences of Physical Problems Involving Differential Equations, with an Application to the Stresses in a Masonry Dam, Philosophical Transactions of the Royal Society of London A: Mathematical, Physical and Engineering Sciences, 210, 307–357, doi:10.1098/rsta.1911.0009, 1911.

Richardson, L. F.: Weather prediction by numerical process, Cambridge University Press, 1922.

Robert, A.: A stable numerical integration scheme for the primitive meteorological equations, Atmosphere-Ocean, 19, 35–46, 1981.

Robert, A. J.: The integration of a low order spectral form of the primitive meteorological equations(Spherical harmonics integration of low order spectral form of primitive meteorological equations), Journal of the Meteorological Society of Japan, 44, 237–245, 1966.

Sadourny, R.: The dynamics of finite-difference models of the shallow-water equations, Journal of the Atmospheric Sciences, 32, 680–689, 1975.

Sawyer, J.: Large scale vertical motion in the atmosphere, Quart. J. Roy. Meteor. Soc, 75, 185–188, 1949.

Simmons, A. J. and Burridge, D. M.: An Energy and Angular-Momentum Conserving Vertical Finite-Difference Scheme and Hybrid Vertical Coordinates, Monthly Weather Review, 109, 758–766, 1981.

Smagorinsky, J.: General circulation experiments with the primitive equations: I. the basic experiment*, Monthly Weather Review, 91, 99–164, 1963.

Smolarkiewicz, P. K., Deconinck, W., Hamrud, M., Kühnlein, C., Mozdzynski, G., Szmelter, J., and Wedi, N. P.: A finite-volume module for simulating global all-scale atmospheric flows, Journal of Computational Physics, 314, 287–304, 2016.

Sommerfeld, A.: Partial differential equations in physics, Academic press, 1949.

Staniforth, A. and Côté, J.: Semi-Lagrangian integration schemes for atmospheric models - A review, Monthly Weather Review, 119, 2206–2223, 1991.

Tallqvist, H.: Grunderna af teorin för sferiska funktioner jämte användningar inom fysiken, Helios, Helsingfors, 1905.

Wiin-Nielsen, A.: The birth of numerical weather prediction, Tellus A, 43, 36–52, 1991.

Williams, P. D.: A proposed modification to the Robert-Asselin time filter, Monthly Weather Review, 137, 2538–2546, 2009.